U0127761

教师用书

Richard Hallows（英） Martin Lisboa（英） Mark Unwin（英） 编著 咸珊珊 译

捷进雅思高级教程

IELTS Express Upper Intermediate

Teacher's Guide

外语教学与研究出版社
FOREIGN LANGUAGE TEACHING AND RESEARCH PRESS
北京 BEIJING

京权图字：01－2006－7591

© First published by Thomson ELT, a division of Thomson Learning, United States of America. All Rights Reserved.

Reprinted for People's Republic of China by Thomson Asia Pte Ltd and FLTRP under the authorization of Thomson Learning. No part of this book may be reproduced in any form without the express written permission of Thomson Learning and FLTRP.

This edition is for sale in the mainland of China only, excluding Hong Kong SAR, Macao SAR and Taiwan Province, and may not be bought for export therefrom. 只限中华人民共和国境内销售，不包括香港、澳门特别行政区及台湾省。不得出口。

978－7－5600－6467－3

图书在版编目(CIP)数据

捷进雅思高级教程：教师用书 = IELTS Express Upper Intermediate Teacher's Guide / (英)哈洛斯(Hallows, R.)等编著；咸珊珊译．—北京：外语教学与研究出版社，2007.3

ISBN 978－7－5600－6467－3

Ⅰ．捷…　Ⅱ．①哈…②咸…　Ⅲ．英语—高等教育—教学参考资料　Ⅳ．H31

中国版本图书馆 CIP 数据核字 (2007) 第 030854 号

出 版 人：李朋义
责任编辑：韩晓岚
封面设计：刘　冬
出版发行：外语教学与研究出版社
社　　址：北京市西三环北路 19 号 (100089)
网　　址：http://www.fltrp.com
印　　刷：北京密云红光印刷厂
开　　本：889×1194　1/16
印　　张：8
版　　次：2007 年 3 月第 1 版　2007 年 3 月第 1 次印刷
书　　号：ISBN 978－7－5600－6467－3
定　　价：19.90 元 (含 DVD 光盘 1 张)

*　*　*

如有印刷、装订质量问题出版社负责调换
制售盗版必究　举报查实奖励
版权保护办公室举报电话：(010)88817519

Contents

什么是雅思考试?

雅思考试由剑桥大学考试委员会（Cambridge ESOL）、英国文化委员会及澳大利亚教育国际开发署联合开发。英国、澳大利亚、爱尔兰、新西兰、加拿大、南非等国家的绝大多数大学和继续教育学院以及美国的很多院校都认可这一考试。该考试也得到了职业机构、移民局和其他政府机构的认可。

雅思考试分为两种形式——学术类和普通类。两种形式的听力和口语测试内容均相同，阅读和写作测试内容不同。学术类测试适用于计划进入大学或研究生院学习的考生，而普通类测试则适用于计划接受非学术类培训、工作或移民的考生。

要想获得更多的考试信息，可访问雅思考试网站：www.ielts.org。

关于雅思考试的4个部分及其内容，请参见《捷进雅思中级教程（学生用书）》第3页至第4页上的“雅思考试概览”。

雅思考试的评分标准以及考官的要求

雅思考试的每一部分都以0－9分来计分。考生会拿到雅思考试各部分的单项分数和一个总分，该总分是4个独立部分的平均分。

在听力和阅读测试中，考生每答对一题得1分，然后该项考试的分数将被转换成9分制的分数。分数的转换方法每场考试各不相同，而且都是保密的。考试的分数可能是整数分（如6分），也可能是分数分（如6.5分）。

写作和口语测试由受过培训的考官来评分。以前，这两部分的分数都是以整数计算。从2007年1月起，这两部分也将实行分数分制。写作和口语测试基本上从以下几个方面来评价考生。

写作

写作Task 1（学术类和普通类）

- **任务完成情况**：作文的完成情况，字数不少于150词。
- **连贯性和一致性**：表达信息的准确性与语言的通顺与否。
- **词汇范围**：考生使用的词汇量、准确度和适用性。
- **语法范围及使用的准确与否**：考生正确使用各种不同句式的能力。

写作Task 2（学术类和普通类）除了“连贯性和一致性”、“词汇范围”和“语法范围及使用的准确与否”以外，还包括：

- **任务完成情况**：考查考生构思并阐述某一论点的能力和论据是否充分。论据可来源于考生的亲身经历。字数不少于250词。

口语

- **流利性和连贯性**：考官将语速和连续性作为语言流利的两个主要评分标准。句子的逻辑顺序、讨论阶段的划分明确与否和衔接手段（如连词等）的使用是连贯性的重要标准。
- **词汇范围**：标准同雅思写作测试中的“词汇范围”这一要求。此外，考官希望考生在遇到生词时，能使用其他的词来代替，自信地克服词汇障碍。他们很重视考生这方面的能力。
- **语法范围及使用的准确与否**：考官将句子的长度和复杂程度、从句的使用与否、句式的变换以及考生利用句式变换来解释重要信息的能力作为衡量考生语法能力的主要标准。语法使用的准确与否是以“一定量的语言中语法错误的数量和由于错误而产生的沟通困难的程度”来衡量的。
- **发音**：这是以语言能否被听懂、第一语言的干扰程度和给听者带来的压力大小来衡量的。

要想获得更多的考试信息，请访问雅思考试网站：www.ielts.org。

每部分考试中考生遇到的困难

考生可能在雅思考试的各个阶段遇到不同的困难。"捷进雅思系列教程"对这些困难都进行了透彻的分析，详细列举了多种应对策略，并提供了相应的练习。考生在考试中通常遇到的困难有以下几种：

听力

每部分的录音只放一遍，考生需要边听边写。说话人可能带有口音，录音中出现的生词还可能较多。这样就会大大增加考试难度。例如，在听力的第4部分（学术独白），考生可能很难跟上说话人的思路。考生还要能够听出文中出现的数字。另外，虽然只是听力测试，但考生也要注意将答案拼写正确。

阅读（学术类和普通类）

多数考生都觉得这部分非常具有挑战性。在学术类考试中，要求考生在短短60分钟的时间内阅读3篇很长的文章，总字数高达2,700词；同时还要回答40道题。此外，文章的内容通常是许多考生所不熟悉的。由于内容晦涩、生词量大，即便给出某些生词的释义，考生也不容易弄明白。此外，考生还可能在一些不太常见的题型（例如，标注流程图）上，因经验不足而失分。

写作

Task 1（学术类）

考生除了要很好地掌握语言技巧，还需要会分析、筛选和排列数据。要能够看出总体趋势，辨认出重要的信息和总体趋势的特例。

Task 1（普通类）

考生很可能不知道在信件写作中该如何取舍信息。其他可能出现问题的地方包括：信件结构、内容安排和语气的轻重。

Task 2（学术类和普通类）

可能会要求考生就一个他/她可能从没有仔细思考过的、根本没有任何想法的主题写一篇论述详尽并且论据充分的议论文！有些考生可能对论证式的写作文体不太熟悉，在术语的使用上也会出现问题。

口语

口语测试的第一部分通常不会太难，但第二部分就要求考生连续陈述1分钟以上的时间；这对许多考生来说是个艰巨的任务。而最后一部分谈论的话题则可能会非常抽象。很多考生会觉得自己的想法和词汇量都非常有限。

"捷进雅思系列教程"如何帮助您的学生？

"捷进雅思系列教程"全方位的课程内容使之成为雅思考试全面有效的备考用书。重要而有效的雅思考试技巧分布在学生用书的各个单元，每个单元会提供两个技巧的讲解。如果考生在听、说、读、写4个方面都需要提高，教师可以按单元的编排顺序使用学生用书：先学第1单元（阅读和口语），再学第2单元（听力和写作），依此类推。或者教师可以根据情况灵活使用本教程，集中讲解学生需要强化训练的一到两个部分。此外，学生用书除了着眼于应试技巧，还着重培养考生的语言技能。

阅读

"捷进雅思系列教程"涵盖了所有主要的学术类和普通类阅读题型。尽管两种形式的考试试题不同，但因为题型是一样的，所以书中并未分别讲解。

写作

第 2 单元和第 6 单元专门讲解学术类Task 1的写作。第4单元和第8单元专门讲解学术类和普通类Task 2的写作。在学生用书的最后，还有3个针对普通类考试的写作单元——两个有关Task 1的单元，一个有关Task 2的单元。

听力

“捷进雅思系列教程”涵盖了雅思听力考试的4个部分，介绍了雅思听力测试的主要题型。所有的录音材料都包含在《捷进雅思中级教程（学生用书）》的配套CD光盘中。

口语

“捷进雅思系列教程”的口语部分涵盖了雅思口语测试的3个部分，对每一部分都作出了详细的分析。另外，《捷进雅思中级教程（教师用书）》的配套口语DVD光盘向考生展示了一个仿真的雅思口试场景。DVD光盘中附有考官的评价，不仅深入剖析口语测试中的各个部分，而且还专门结合学生用书口语单元中练习过的技巧对考生的表现加以点拨。请使用教师用书中第117页至第121页可复印的练习材料来配合DVD光盘的使用。关于如何在课堂上使用DVD光盘的说明，请参见教师用书第109页至第121页。

学生用书的单元结构使教师能灵活安排课程进度，并帮助学生把更多的精力集中在强化薄弱环节上。《捷进雅思中级教程（强化训练）》提供了更多的技巧训练和模拟试题，同时还可以扩大考生的词汇量并帮助考生集中精力学习重要的语法结构。

“捷进雅思系列教程”的产品包括哪些？

《捷进雅思中级教程》（适合英语基础比较薄弱的考生）和《捷进雅思高级教程》（适合旨在取得雅思高分的考生）都各自包含一套完整的配套产品。这些产品包括：

学生用书

每册学生用书首先介绍雅思考试，然后是按常见的雅思话题设置相关的8个单元。中级学生用书的后面是针对普通类考试编写的一组单元。中级和高级的学生用书都提供一套完整的模拟试题，并附有答案、写作范文、听力与口语练习的全部录音文本。为了保证测试的有效性，模拟试题的答案只附在教师用书中。教师用书还附有模拟试题的阅读和听力测试需要用到的答题纸。在学生用书中第106页至第109页上，附有关于写作测试的实用表达和参考句式，便于考生查找使用。

CD光盘

CD光盘除了包括每个口语部分范文的录音外，还包括听力单元中的练习题和模拟试题中听力测试的录音。

强化训练

强化训练是学生用书的补充，通过扩大词汇量和掌握更多的语法知识来帮助考生提高语言水平，并提供更多的解题技巧和模拟试题。本书适合课堂教学和自学，所附CD光盘包含听力和口语测试的录音材料。

教师用书

教师用书提供了详细的课程讲解，指导教师如何进行各项练习及如何处理可能出现的问题。此外，本书还附有如何帮助基础较弱的考生（标有*Support*处）和进一步帮助基础较好的学生（标有*Challenge*处）的建议。这些教学建议对不太熟悉雅思考试的教师和经验丰富的教师都很有帮助。

口语 DVD 光盘

DVD 光盘向学生展示了一个仿真的雅思口试场景，附有考官的评价，考官会说明口试中的各个部分，并结合学生用书口语单元中的技巧对考生的表现加以评价。请使用教师用书中第117页至第121页可复印的学生练习材料来配合DVD 光盘的使用。关于如何在课堂上使用DVD 光盘的说明，也请参见教师用书第117页至第121页。

如何在课堂上使用“捷进雅思系列教程”？

“捷进雅思系列教程”设计灵活，适用于不同长度和/或水平的各类雅思培训课程。

30课时的课程

每册学生用书都提供了30－40小时的课时。这对为期一周的雅思备考课程是很有效的。每个单元大约需要4小时，一本学生用书的8个单元将提供32小时左右的教学材料。加上口语测试，一套模拟试题需要2.5－3小时。如果打算对学生逐个进行口试，每个学生大约需要15分钟。其他进行口试的方法请参见教师用书第94页。

每个级别60课时的课程

配合使用强化训练和口语 DVD 光盘，课时会多出30－40小时。

两个级别60课时的课程

在两个级别的强化备考课程中，教师可以在课堂上使用学生用书，学生在家里完成强化训练中的习题。

多于90课时的课程

对于时间比较充裕的雅思备考课程，教师可以使用“捷进雅思系列教程”中两个级别的全部产品。本系列教程至少提供了90小时的教学材料，完全可以满足一个为期一学年、全方位的雅思备考课程的需要。

如何教授“捷进雅思系列教程”？

学生用书的每个单元都包括雅思测试的两个部分。第一部分考查的是输入性技巧：阅读或听力测试；第二部分考查的是输出性技巧：口语或写作测试。每个部分需要2小时左右的授课时间。教师应该鼓励考生提高做题速度，在考生做题时给出一个时间限制。教师用书还提供进一步的拓展练习，使用这些练习可以延长授课时间。

学生用书的内容构成

每单元的各部分（阅读、口语、听力和写作）都经过精心编排，分别介绍并强化重要的考试技巧，指导考生如何使用这些技巧来答题。最后，在每单元的结尾部分，还给学生提供了模拟试题来运用新学会的考试技巧。

一个单元通常包括下列部分：

本单元的训练内容：每一页的上方会标明本单元涉及的训练内容。

图片：每单元都配有与主题相关的图片，这些图片给考生提供了一个清晰的视觉焦点。

Introduction：通过讨论问题来介绍本部分的主题。

IN THE EXAM：该部分是学生用书的一个重要内容，帮助考生将每部分所练习的技巧与整个考试联系起来，详细描述考试的各个方面，并对考试中可能出现的一到两个题型作详细的说明。

Express tip：该部分为应试提供实用的策略及技巧。

For this task：该部分出现在练习题的前面。它为某个特定的考试题型提供相应的应试技巧，指导学生如何将新学到的技巧应用到稍后的练习题中。

强化技巧练习：介绍并练习重要的雅思应试技巧，例如略读和查读、判断总体趋势、识别释义等。

练习题：每个部分都以一个练习题结束，使考生有机会来使用新学会的应试技巧。

教师用书的内容构成

学生用书的每个部分在教师用书中都有一套相应的课程讲解和教学建议。教师用书包括下列部分：

本单元的训练内容：每一页的上方都标明本单元涉及的训练内容。

Section aims：说明每部分要达到的主要目标。

Aims：详细介绍每个练习的目的。

ANSWER KEY：解释了为什么一个答案是正确的，而其他的答案是不正确的。

Listening scripts：为了使用方便，除了学生用书附有听力录音文本外，教师用书也提供了听力测试和口语测试的录音文本。

Support：对如何帮助那些觉得某个练习比较困难的考生提供建议。

Challenge：对如何进一步帮助那些觉得某个练习相对容易的考生进行进一步练习提供建议。

Extension：如果教师认为考生能通过进一步的练习得到提高，或者发现课堂时间充裕、需要额外的训练材料，可以使用这部分扩展练习题。

教学建议：循序渐进地指导教师如何教授某个特定的练习——介绍主题或练习题、提前教生词、将考生分组、预测问题、限时练习、分阶段练习听力、使用学生用书中的各种提示、引出考生的反馈和检查答案等。

关于如何使用模拟试题及相关的评分标准，请参见教师用书第 92 页至第94页。

1

Leisure Activities

READING

Section aims:

- To teach students a procedure to follow when first approaching a passage.
- To introduce and practise the skills of skimming to understand main ideas and scanning to locate information within the text.
- To practise a number of task types – matching headings to paragraphs; summary completion; short-answer questions.

1 Introduction

Aims: To introduce the topic of leisure activities and to act as a lead-in to the reading text about stress and holidays.

To introduce some of the vocabulary that arises later in the unit.

A
- Direct students' attention to the photo and ask them questions to orientate them to the topic such as: *Where do you think this photo was taken? Has anyone been to somewhere similar? What can you do on this kind of beach holiday?*
- Put students into pairs and have them discuss the questions with their partner.

B
- Explain to students that the words in the language box will appear in the reading passage. Have them work individually or in pairs to group the words into the three categories.

ANSWER KEY

calm: relaxed, utterly unstressed

quite stressed: a little frustrated, pretty anxious, somewhat nervous, slightly edgy

stressed: really stressed out, incredibly uptight

Support

- Check that students understand the meaning of the words *calm, stressed* and *worried* by asking them to plot the three words on the following scale (drawn on the blackboard):

☺ ____ ☹

(*calm* = ☺, *stressed* = ☹ and *worried* would be approx ½ the way along the scale near ☹)

Challenge

- To practise using intensifiers, point out the words *somewhat, really, a little, pretty, slightly, incredibly* and *utterly* from the language box and ask students the following question: *What function do these words have?* Elicit that they make the adjective that follows them more, or less intense. In fact, they are known grammatically as 'intensifiers' or 'submodifiers'.
- Ask students to rank these words (1–7) in order of intensity from weak to strong.

ANSWER KEY

1/2 a little, slightly; **3** somewhat; **4** pretty; **5** really;
6/7 incredibly, utterly

- Ask students if they know any other intensifiers. **Suggested answers:** *quite, very, rather, totally, reasonably*
- Ask students to add these words to the list above and rank them.

ANSWER KEY

1/2 a little, slightly; **3/4** somewhat/reasonably*; **5** quite;
6/7 pretty, rather; **8/9** really/very; **10** totally;
11/12 incredibly/utterly

Note: *Somewhat* is usually followed by an adjective with a negative connotation; *reasonably* is followed by one with a positive meaning (e.g. *calm*).

C
- Ask students to read through the instructions and then have them close their eyes for a minute so they can imagine different situations where they have felt these emotions. Give them a minute or so to take down a few notes. Then put them into pairs to compare and discuss.

IN THE EXAM

- Use the information in the *IN THE EXAM* box to introduce the IELTS Reading module. Either have the students read the information to themselves, or read it aloud while students follow. After students have read the information, ask them to close their books, then ask questions to check their comprehension, for example: *How long is the reading paper? How many questions are there?* etc.

2 Approaching the text

Aims: To teach a procedure students can follow when they first approach an IELTS reading passage.

To show how students can predict the content of the passage by looking first at the title, photo, caption and first paragraph.

To show students how an initial skim read can achieve a general understanding of the text.

A ▸ Explain to students the standard procedure for first approaching IELTS reading passages, i.e. before starting to read the passage, they should look first at the title, subtitle, any photos or illustrations and the first paragraph. Explain to them that IELTS is all about speed reading and that they will need to read in a particular way which will no doubt be different to the way they normally read in their own language. Making initial predictions about the content of the passage will help them to read and understand the passage more quickly.

▸ Ask students to read the instructions and do the task.

ANSWER KEY

Holidays can be stressful.

B ▸ Ask students to skim the passage to decide the main theme. Emphasise to students that you want them to *skim* the text rather than read it in detail. To ensure that they only skim, give students only two or three minutes to do the task. Refer students to the *express tip* box on page 10 for a brief definition of *skimming*. Ask the class whether their predictions in the previous section were correct.

ANSWER KEY

The article is aimed at a general readership. You might see this text in a newspaper.

3 Skimming for main ideas

Aims: To teach students how to understand the main ideas of the passage by focusing on the topic sentence of each paragraph as they skim the passage a second time.

To teach students how to identify the main idea of each paragraph by separating it from supporting information.

express tip

▸ Read out the advice in the box while students follow in their books.

A ▸ Ask students to read the instructions to the task. Explain that understanding the main ideas of the passage will enable students to locate answers to questions much more quickly – it provides them with what is sometimes referred to as a 'map of the text'. Explain that each paragraph should have a topic sentence/main idea: this text with eight paragraphs should have eight main ideas.

▸ Ask students to read through the instructions and then do the task.

ANSWER KEY

Paragraph E: 'Now we have a formula to calculate the amount of holiday time needed to recover from the stress of preparing for what should be our annual period of rest and recuperation.'

Paragraph F: 'Today's pressurised lifestyles mean that going on holiday is a lot more stressful.'

Paragraph G: 'So according to Ms Quilliam, the secret to a stress-free holiday is planning and having realistic expectations.'

Paragraph H: 'The results of a survey commissioned by Lloyds TSB bank and carried out by Ms Quilliam, have confirmed her theories; today's holidaymakers really are stressed out.'

B ▸ Explain to students that identifying topic sentences is the first step to identifying the main ideas of the passage. The second step is to summarise each paragraph in their own words. This will require particular attention to the topic sentence, but will also draw on points in the rest of the paragraph. By summarising the paragraph students will further their understanding of each main idea.

▸ Ensure students have identified the correct topic sentences for paragraphs E–H before doing this task. As a model, do a paragraph summary for Paragraph E as a class. For the remaining paragraphs, ask students to write summaries individually or in pairs.

▸ Ask students to read through the paragraph summaries and decide which of each pair is correct, *a* or *b*.

ANSWER KEY

1 A a; **B** a; **C** b; **D** b

Answer key

2 Suggested answers:

Paragraph E: 'There is a formula for calculating the number of hours needed on holiday to recover from the stress of organising the holiday.'

Paragraph F: 'Today's complex holiday preparations mean holidays are more stressful than they used to be.'

Paragraph G: 'If you plan carefully and do not have too high expectations, you can have a stress-free holiday.'

Paragraph H: 'A survey showed that most people get stressed on the run-up to their holiday.'

C

- Ask students to analyse the example of Paragraph B by asking them to match the supporting ideas to the relevant parts of the paragraph in the passage. Supporting ideas include examples, further explanation, supporting evidence, etc. It should be kept in mind, however, that not all non-topic sentences in a paragraph are supporting ideas. Some might be 'bridge sentences', used to connect the paragraph to the main idea of the previous or following paragraph.
- Once students are clear about the difference between main and supporting ideas, put them into pairs to find the supporting ideas for the remaining paragraphs. Make sure students write their answers as 'mini summaries' as in the model, and do not simply copy out chunks of text from the passage.

Answer key

Suggested answers:

Paragraph A: Examples of travel arrangements.

Paragraph C: Body needs time to regain balance.

Paragraph D: Examples of psychological problems.

Paragraph E: Explanation of the formula.

Paragraph F: Examples of today's complex travel arrangements.

Paragraph G: Examples of good planning.

Paragraph H: Detailed breakdown of survey results.

Support

- If students are not able to move beyond copying out sections of the text to paraphrasing in their own words, then work through each paragraph as a class. Ask what the type of supporting evidence is, for example: *Looking at Sentence 2 of Paragraph A. Is the supporting evidence here a list of facts, examples or an explanation?*
- Ask students to summarise the information in their own words. Tell students to start their summary with the words *Examples of ...* Finally, ask a good student to write his/her answer on the blackboard and then proceed to analyse and correct it together with the class. Elicit answers as much as possible by asking questions: *Is this the correct spelling here? Is the grammar correct here? Are all the words necessary? Is any information missing?* etc.

Extension: *Analysing a newspaper article*

Aim: *To provide further practice in identifying topic sentences, and main and supporting ideas.*

Preparation

- Before class, find a short newspaper article, consisting of eight to ten paragraphs on the topic of leisure activities (sports/hobbies/entertainment/holidays, etc) or stress (work stress/causes of stress/dealing with stress, etc). Choose a well-written text, with one main idea per paragraph. Make enough copies of the article in order to provide at least one copy per pair or small group of students.

Procedure

- Divide the class into pairs or small groups and give one copy of the article to each pair or group. Have each pair or group analyse the paragraphs in the article to identify topic sentences, and then the main idea in each paragraph as well as any supporting ideas. You may wish to do the first paragraph as a class. If time is short, assign only a few paragraphs to each group.
- When students are finished, elicit answers as a class and discuss.

D

- Ask students to read the instructions to the task, then do the task individually. Put students into pairs to compare and explain their answers. Ask one student to come to the front of the class to explain his/her answer using the blackboard as an aid.

Answer key

Example answer:

$$\frac{P(20\text{ hours preparation}) \times S(4\text{: slightly edgy})}{H(10\text{: medium stress})} = R\ (8\text{ hours rest before enjoying the holiday})$$

4 Scanning for keywords

Aims: To practise identifying keywords within exam questions.

To introduce the skill of scanning to locate specific information in the text using keywords from the questions.

To teach students to recognise paraphrase and synonyms.

A
- Make sure students understand the terms *scanning, keyword, paraphrase* and *synonym*. Suggested definitions are:
  Scanning: running your eyes quickly over the text to find specific information
  Keyword searching: scanning for keywords. Students are identifying the keywords in the questions and searching through the text to find keyword matches.
  Paraphrase/Synonym: word or words which have a similar meaning. A synonym refers to a single word (e.g. a synonym for *tutor* is *teacher*). Paraphrase refers to more than one word – an expression or sentence: *This PC has crashed* is a paraphrase of *This computer has stopped working*. Synonyms and paraphrase are important for keyword searching as there is often not an exact match between the keyword in the exam question and the target word in the text – often students will need to match the keyword with a synonym or paraphrase instead.
- Ask students to read the sentences, paying attention to the underlined words. Ask them to think of synonyms or paraphrase for the underlined words. Next have them scan the passage to find the section of text each sentence is taken from. When they have found the relevant sections in the passage, ask students to underline the appropriate paraphrase (or exact keyword match) in the passage.
- Explain that keyword searching by scanning the text tells you where the answer is; it doesn't tell you what the answer is.

ANSWER KEY
Underlined words: 1 fantasise, everyday grind; **2** expert; **3** more than a quarter

express tip
- Read out the advice in the box while students follow in their books. Demonstrate by physically acting out the difference between reading and skimming. Exaggerate movements of the head from left to right with the book in front of you to show what reading is, i.e. reading each line of the text from left to right. Then show students what skimming is by placing your index finger on the page and pulling it vertically down – your eyes moves *down the page, not across*. An alternative method of skimming is to sweep your eyes diagonally across the page. This is perhaps best illustrated on the blackboard. Students who are unfamiliar with skimming will find this 'unnatural' at first so you'll need to persist. Like any skill this may seem very awkward and difficult in the beginning but will become second nature with continued practice.

B
- Now direct students to read that specific part of the passage in detail to complete the gaps in the sentences.

ANSWER KEY
Gap completion: 1 (long) break; **2** body language; **3** recover

C
- Remind students that the key to keyword searching is knowing what word to search for – in the previous exercise (4A) the questions already had the keywords correctly identified and underlined. This exercise is more difficult as students first have to underline the keyword before scanning and answering the question. Ask students to work individually to identify and underline the correct keyword(s) in each sentence and then compare their answers.
- After students have underlined the correct keyword and identified the appropriate match in the passage, they are ready to complete the gaps in the sentences.

ANSWER KEY
Keywords: 1 swimming costume, formal clothes; **2** psychological symptoms; **3** research
Gap completion: 1 reading material; **2** anxiety, irritability, mild depression; **3** Lloyds TSB Bank

5 Matching headings to paragraphs

Aims: To introduce and practise the exam task type: matching headings to paragraphs.

To consolidate skills introduced in this unit.

for this task

- Read through the *for this task* box aloud and ask students to follow in their books. Point out that they will already have done all the preparatory work for this question type in the preceding skills sections by having already identified the topic sentence, summarised each paragraph in their own words and distinguished the main idea from the supporting info and subsidiary ideas.

EXAM PRACTICE

Questions 1–8

- Ask students to work alone to answer Questions 1–8.
- On completion, ask students to compare their answers with a partner, or check answers as a class.

ANSWER KEY

1 Answer: iii
Note 'First there's the flight to arrange, then the hotel or villa to book … the beach outfit, evening wear and reading material.'

2 Answer: ii
Note '… making our dream holiday a reality is what makes the stress really kick in …'

3 Answer: x
Note 'stress … creates hormonal changes in the body, including mood alteration.'

4 Answer: ix
Note '… if we get stressed before the holiday, we may not be able to relax … often for several days after our arrival.'

5 Answer: vi
Note 'Now we have a formula … The formula … is calculated thus …'

6 Answer: vii
Note '… taking a holiday is no longer a question of catching a bus to the nearest seaside resort … Today's pressurised lifestyles mean …'

7 Answer: v
Note '… the secret to a stress-free holiday is planning and having realistic expectations.'

8 Answer: i
Note 'The results of a survey … have confirmed her theories. More than 83 per cent …'

6 Summary completion

Aim: To introduce and practise summary completion questions.

for this task

- Read through the *for this task* box and ask students to follow in their books. Remind students of their previous skills work identifying main ideas and the preceding exam question type – matching headings to paragraphs. Point out that they should be using their understanding of the organisation of the text in order to quickly locate the part of the text that relates to the summary.

EXAM PRACTICE

Questions 9–13

- Ask students to answer the questions individually and compare their answers with a partner on completion.
- When eliciting answers from students as a class, focus on keyword paraphrase as a means of navigating through the passage and summary. For example, the synonym in the passage for the keyword *mathematical equation* is *formula*. Identifying this word quickly is the important stepping stone to finding the answer to Question 9.

ANSWER KEY

9 Answer: devised
Note 'The formula, devised by Ms Quilliam …'

10 Answer: pressurised lifestyles
Note 'Today's pressurised lifestyles mean that going on holiday is a lot more stressful.'

11 Answer: realistic expectations
Note '… the secret to a stress-free holiday is planning and having realistic expectations.'

12 Answer: (any) unfinished work
Note 'This means concluding any unfinished work in the office …'

13 Answer: tell their bosses
Note 'Amazingly, a tiny minority of travellers (a little over 1 per cent) actually forgot to tell their bosses they were going on holiday!'

7 Short-answer questions

Aim: To introduce and provide practice of short-answer questions.

for this task

- Read through the *for this task* box and ask students to follow in their books. As with summary completion questions, point out to students the link with the preceding skills sections, i.e. understanding of text organisation/main ideas to locate the relevant part of passage and using keyword matching between questions and passage.

EXAM PRACTICE

Questions 14–17

- Ask students to answer the questions individually and compare their answers with a partner on completion.

ANSWER KEY

14 hormonal
Note '... stress such as that caused by trying to arrange a holiday, creates hormonal changes in the body ...'

15 a formula
Note 'Now we have a formula to calculate the amount of holiday time needed to recover from the stress ...'

16 bucket and spade
Note '... taking a holiday is no longer a question of catching a bus to the nearest seaside resort with your bucket and spade.'

17 severely stressed
Note 'More than 83 per cent of people surveyed admitted to getting 'severely stressed' in the run-up to their holiday.'

Extension: *Reverse reading comprehension*

Aim: *To provide extended practice of short-answer questions with authentic material.*

Preparation

- Download from the Internet a number of appropriate magazine or newspaper articles, one for every two or three students. Make sure the articles are not too long (8–10 paragraphs) and are on an interesting topic for students, linked to the theme of leisure activities.

Procedure

- Put students into pairs (or groups of three) and hand them a magazine or newspaper article (each pair or group will have a different article).
- Ask students to read through the articles in their pairs or groups using appropriate IELTS reading techniques, (see sections *Approaching the text* and *Skimming for main ideas*). Answer any vocabulary questions.
- Once students have read and understood the text, have them write five short-answer comprehension questions. Ask students to write down the questions on a separate piece of paper with the answers on the back. Go around checking that the questions are meaningful (i.e. they have a definite answer) and are grammatically correct.
- Ask students to pass their article and accompanying questions to another pair/small group, who will need to read the article (using IELTS reading techniques) and answer the questions (scanning for keywords). Groups continue passing the articles to new pairs or groups until all the students have seen all the articles or until enthusiasm wanes. This activity should take 30–60 minutes depending on the size of class.

Section aims:

- To introduce the types of questions students might encounter in Parts 1 and 2 of the Speaking exam, and give them the opportunity to develop their own answers.
- To promote fluency by encouraging students to extend their answers and give additional information in Part 1 of the exam.
- To demonstrate how candidates can organise their ideas in Part 2 of the Speaking exam.
- To practise Speaking Part 1: Introduction and interview and Part 2: Individual long turn.

1 Introduction

Aims: To introduce students to the types of everyday questions they are likely to be asked in Part 1 of the Speaking exam.

To help students get to know each other.

- Explain that students are going to find out something about their classmates. Elicit some typical questions that you might ask someone when you first meet them. You should try to elicit questions belonging to the five categories in Section 2. You could write up the categories and get students to supply possible questions. When you elicit questions, get students to give you a likely 'follow-up' question, e.g. *Where are you from? How long have you been living there? Do you have any brothers or sisters? Do you get on with them?*
- Ask students to try to find out a little about another student. Encourage them to use questions from a range of categories as well as follow-up questions. To extend the activity, ask them to change partners and speak with another student.

Support

- If students need some help coming up with questions, write some typical questions jumbled up on the whiteboard/on a photocopy and ask students to put them in the right order. Some typical questions might be: *What's your name? Where do you come from? How long have you lived there? Do you like living there? Have you got any brothers or sisters? Do you get on with them? Have you got a job? What are you studying at the moment? What would you like to do after your studies?*

IN THE EXAM

- Read the information in the *IN THE EXAM* box while students follow in their books. Draw attention in particular to the marking criteria. Explain the marking criteria: *fluency* (students should try to avoid long pauses and hesitations); *coherence* (the ideas should be easy to follow and have a logical flow to them); *lexical resource* (students should use a range of words accurately, including some lower frequency words); *grammatical range and accuracy* (students should accurately demonstrate a use of a wide range of grammar structures); *pronunciation* (it is not important for students to sound like native speakers, but comprehension should not be impaired by poor pronunciation).

2 Giving personal information

Aims: To give students a chance to hear some candidates answering typical Part 1 questions.

To practise answering typical Part 1 questions.

- Tell the class that they are going to hear seven exam candidates answering the examiner's questions. Explain that they won't hear the examiner's actual questions; they will only hear the answers. Students should predict what questions the candidates were asked and write down their predictions. This may be done in pairs.
- Play the recording. If necessary, pause the recording after each speaker to give students time to write. Ask students to compare their answers, then check as a class.
- Throughout the Speaking sections in the book, you should try to maximise the opportunities students get to practise the different parts of the exam. After giving feedback, put the students into pairs to role-play the exam. One student should act out the role of examiner; the other should play the role of student.

1.1 LISTENING SCRIPT

1

Candidate 1: Yeah, I have just one brother. He has two children, er ... sons, Juan and José. That makes me the uncle! I really like playing with his children, but it's always good to give them back to him at the end of the day! When I'm older I'd like to have some children myself, maybe two boys and a girl. I think that having children ...

2

Candidate 2: I really enjoy speaking it, and I like it when I learn a new word which is very useful or is funny in some way. But it's very hard work and sometimes I feel like I don't get better, er ... I'm not making a lot of progress. I prefer studying economics, which is the subject I'm hoping to study when I go to university next year. You know, it's interesting to learn about ...

3

Candidate 3: Not really, I'm not very sporty. I prefer to spend time playing on my computer. In fact, I really enjoy video games. I got a new game recently – it's really great. You're a criminal and you have to drive your car really fast about a city, and then sometimes escape from the police. It's very exciting. Sometimes I play with my friends, but it's quite difficult to ...

4

Candidate 4: I work in McDonald's. It's a part-time job. It's very good because it gives me time to study and I meet people to practise my speaking, you know, talking with my colleagues ...

5

Candidate 5: I don't know exactly what I want to do, but when I'm older I'd like to work for an international agency – I think it would be a good job, interesting with lots of opportunity for travelling – but right now I have to study hard for the IELTS, to go to university, to get my degree ...

6

Candidate 6: I'm from Sao Paolo in Brazil. I've travelled to different cities with my job. I once went to Tokyo for a meeting. That was an amazing experience, so different from Sao Paolo. The people are ...

7

Candidate 7: I really enjoy science subjects: chemistry, physics and biology. My favourite was physics at school, but I'm planning to study computer science in the future.

ANSWER KEY

1 Do you have any brothers or sisters?
2 Do you enjoy studying English?
3 Do you play any sport?
4 Have you got a job?
5 What do you plan to do after you finish your studies?
6 Where are you from?
7 What's your favourite subject?

3 Providing additional information

Aims: To show how students can extend their answers to questions in Part 1 of the exam and sound more fluent.

To provide further speaking practice.

A ▸ Play the candidates' responses through a second time, this time having students listen for the extra information that they give to extend their answers. Ask students to make notes to complete the sentence beginnings in 1–7. Stress that it is not necessary to write down the answers word for word, but rather to get the main idea.

ANSWER KEY

1 I really like playing with his children, but it's always good to give them back to him at the end of the day! When I'm older I'd like to have some children myself, maybe two boys and a girl. I think that having children ...
2 But it's very hard work and sometimes I feel like I don't get better, er ... I'm not making a lot of progress. I prefer studying economics, which is the subject I'm hoping to study when I go to university next year. You know, it's interesting to learn about ...
3 I got a new game recently – it's really great. You're a criminal and you have to drive your car really fast about a city, and then sometimes escape from the police. It's very exciting. Sometimes I play with my friends, but it's quite difficult to ...
4 It's very good because it gives me time to study and I meet people to practise my speaking, you know, talking with my colleagues ...
5 I think it would be a good job, interesting with lots of opportunity for travelling – but right now I have to study hard for the IELTS, to go to university, to get my degree.
6 I've travelled to different cities with my job. I once went to Tokyo for a meeting. That was an amazing experience, so different from Sao Paolo. The people are ...
7 My favourite [subject] was physics at school, but I'm planning to study computer science in the future.

express tip

▸ Read out the advice in the box while students follow in their books. Explain that students shouldn't be afraid of giving wrong answers in any part of the Speaking exam; there are no right or wrong answers. Candidates should try to relax and speak freely without worrying. They should focus more on sounding interesting, saying something that will catch the examiner's interest – this may be one of many interviews that the examiner has conducted today! Giving extended answers is an important way of doing this.

B ▸ Put the students into different pairs and have them repeat the role-play of Part 1 of the exam, using the questions generated in Section 2. You should give students time to think about how they can extend their answers from the responses given in Section A.

Challenge

▸ Ask students to discuss which of the extra information was relevant or irrelevant to the question. Explain that when extending your answer it is important to bear the original question in mind.

4 Introduction and interview

Aim: To provide additional practice of Part 1 of the Speaking exam.

To review the marking criteria for Part 1 of the Speaking exam.

for this task

▸ Read the information in the box as students follow in their books. Draw students' attention again to the marking criteria used in this part of the exam, and make sure that they are clear on what each one means.

EXAM PRACTICE

▸ Put students into pairs (candidate and examiner) and have them role-play this part of the interview before changing roles. Explain that as this part of the exam is not too demanding, they should try to relax and 'settle in' a little.

▸ Instruct the student playing the role of examiner to think of follow-up questions to ask the candidate. Have this student listen specifically for whether the candidate extends their answers and, if so, what extra information do they give?

▸ You could also ask them to think about the criteria explained in the *In the exam* box and give feedback to each other before having a class round up of what candidates did well in and what they need to work on. Be sensitive about exposing students' weaknesses in front of the class.

Support

▸ Your students may find it difficult to come up with follow-up questions. This is not important in the actual exam, as clearly this is the examiner's job. However, it is useful if students are able to anticipate likely questions. Therefore, if students are struggling with this task, you could supply a list on the board for them to choose from.

5 Organising your talk

Aims: To provide exposure to a Part 2 Topic Card.

To provide a model of candidates doing Part 2 of the exam, which students can evaluate.

To get students to think about the organisation of the information they give in their answer.

To help students decide whether a particular piece of information is relevant to their answers.

A ▸ Explain to the class what is required in Part 2 of the Speaking exam and focus students' attention on the Topic Card. Before you look at a sample response to this card, you could ask students to discuss what they might talk about if they were presented with this card. Use the students' ideas as a way of showing students that there is no one correct answer to this task and that they simply need to draw on their own personal experience.

▸ Explain about the time allowed for students to make notes prior to the exam and ask students to look at the notes a candidate has made about a swimming competition.

▸ Have students decide in which order the points should go in order to follow the points on the card and have a coherent flow. Ask them to decide which information is not relevant to the points on the card. Ask students to compare their thoughts in pairs, then elicit answers from the class.

ANSWER KEY

Points 2 and 4 are irrelevant.

B ▸ Tell the class that they will now hear the candidate's response. Play the recording through once, and ask students to check their answers from the previous task. They should also listen for any extra information the candidate gives that is not mentioned in the notes.

ANSWER KEY

The correct order is: 1 I went to see my boyfriend in a swimming competition; **2** I wanted to support him; **3** There were lots of races; **4** He won the breaststroke (competition).

Extra information: he was feeling nervous; had seen (a swimming competition) on TV; he was very slow in the freestyle

1.2 LISTENING SCRIPT

Candidate: OK, let's see. I want to tell you about the time I went to see my boyfriend take part in a swimming competition. It was part of a charity event – you know, making money for good causes. Anyway, why did I go? Well, I wanted to support him, to shout and cheer: it was his first big competition and he was feeling a little bit, er ... nervous. So, it was the first time I had been to such an event. I had seen it on the TV before of course, but when I saw it in real life it was very exciting indeed! I saw many different styles – crawl, this is freestyle, I think ... are they the same? And I saw breaststroke and backstroke and the butterfly. It was fantastic, you know? He was in the breaststroke competition and the freestyle. He was very slow in the freestyle, but the breaststroke competition, he won! It was very exciting and I felt very proud.

C

- Tell the class that they will now hear a second candidate answering the same card. Have students read the candidate's notes. As they listen, students should note which information is included in the response and which has been omitted.
- Play the recording through once, then check the answers.

ANSWER KEY

The speaker forgets to mention that Stephan won second place in the high jump.

Extra information: sports day every year; many students competed in different races; selected to take part by sports teacher; didn't win race because one of the teammates dropped the baton; winners given medals and prizes

Follow-up questions asked: Do you prefer to watch sport or take part? Are you good at football?

1.3 LISTENING SCRIPT

(C = Candidate; E = Examiner)

C: Er, let's see. I want to tell you about when I was at school. We had a sports day every year when many students competed in lots of different races and other things. Why did I go? Well, I used to be quite a fast runner in those days and so I had been selected to take part by a sports teacher of mine. I was chosen to run the 400-metre race where you have to give the baton to another person who is running – do you say 'relay race'? Unfortunately we didn't win as one of my teammates dropped the baton!

So, what did I see? Well, if you weren't taking part in the competition you sat on the grass in the sunshine and enjoyed the other races and events! I saw many of my friends do things like the long jump and the high jump. At the end of the day all the winners were given medals and prizes. I really enjoyed it – it was fun because it was during the summer and I enjoyed not being in class and relaxing in the sunshine watching our mini Olympic Games!

E: Do you prefer to watch sport or take part?

C: I like both, I think. I often play football with my friends and I like to watch it on TV, too.

E: Are you good at football?

C: I'm OK. I don't score many goals, but I enjoy it anyway.

express tip

- Read out the advice in the box while students follow in their books. Although it may seem an obvious thing to say, students often make notes and then forget to look at them while they are speaking. Explain that it's OK to refer to the notes whilst speaking, but not to stop speaking for a long time to read what you've written before starting to speak again. Candidates need to be able to refer to their notes without pausing. Have them practise doing this so they can do it 'smoothly'.

D

- Ask students to make their own notes for the same task. Remind students of some of the ideas that the class came up with when they brainstormed ideas earlier in the section. Explain that while the examiner does not know if the response is based on a true event or not, the response will sound more authentic if it really happened.

- When students have made a set of notes, have them practise their talk with a partner and then change roles.

Support

- It is not important at this stage for students to make their notes within the one-minute time limit. Make students aware of the time limit, but give them two minutes to do this.

Challenge

- If you feel that you want to make the task more demanding, then ask the students to make notes based on the card before they look at the notes in the book. They can then compare their notes with the ones in the book and look for similarities or differences, or look to see whose notes contain useful vocabulary, are easy to follow or are relevant to the prompt card.

6 Individual long turn

Aim: To provide a full Speaking Part 2 practice, where students can consolidate the skills introduced in this unit.

for this task

- Go through the information in the *for this task* box with the class. Draw students' attention to the one-minute time limit for note making. Impress upon students how little time this is: they have to go with their first idea and make notes quickly.

express tip

- Read out the advice in the box while students follow in their books. Explain that the card can be helpful in organising the structure of the talk. Tell students to put their finger on the point they are talking about and keep moving it through the points on the card as they 'move through' the talk.

EXAM PRACTICE

- Put the students into pairs and assign roles of candidate and examiner.
- Have examiners give candidates one minute to make notes to help them respond to the task card.
- Encourage the examiner to assess the performance of the candidate by focusing on the criteria in the *for this task* box, *express tip* boxes as well as the *IN THE EXAM* box on page 13. When the candidate has spoken for two minutes or so, the examiner should interrupt with two or three short questions to bring the turn to a close.
- Have the examiner give feedback to the candidate based on the criteria before changing roles and repeating the activity.
- Monitor closely during this activity and try to include points students need to work on in the class discussion at the end.

IELTS Express Speaking DVD

- If you are using the *Speaking DVD* which accompanies *IELTS Express*, Section 1 – Overview and Section 2 – Part 1 of the DVD relate to the content of this unit.
- It would be helpful to show the Overview before this lesson to give students a general idea of the content and format of the IELTS Speaking exam. The Part 1 section of the DVD could be shown at the end of the lesson to provide a recap and demonstration of the material covered in this unit.

For more information on the *IELTS Express Speaking DVD* and how to integrate it into your lessons, see page 109.

Education

LISTENING

Section aims:
- To introduce the pre-listening skill of anticipating what might be heard, and to highlight the importance of following instructions carefully.
- To provide practice in identifying keywords and paraphrase.
- To introduce and practise Listening Section 1: Non-academic dialogue.
- To practise a number of task types: form completion; multiple-choice questions with single answers.

1 Introduction

Aims: To introduce the topic of education and to introduce some of the vocabulary used later in this unit.

To encourage students to think about the kinds of conversations they might hear later in the unit.

A
- Draw students' attention to the pictures at the top of the page. Ask a student to describe one of the pictures to the class. Ask: *How many people are there? Where are they? How are they dressed? Who are they?* This provides a good opportunity for remedial work on describing things: *There is/are ...; They are wearing ...*, etc.
- Ask the class in what ways the pictures are the same (*both show people in educational facilities*) and in what ways the pictures are different (*the one on the left shows university students; the one on the right shows pupils in a school*).

B
- Ask students to look at the words in the box. Tell them that some of these words relate to school, others relate to university and some relate to both. You could ask students to read the list out, correcting pronunciation and syllable stress as they do so.
- Draw a table on the board with three columns headed: *School, University, Both*. Ask students to copy the table into their notebooks. Now ask students to work in pairs to sort the words into the three different categories, adding the words to the appropriate column. Instruct them to leave a line or two between each word.
- After a few minutes, elicit answers from the class and complete the table on the board, eliciting or providing a definition for each word.

ANSWER KEY

School: report, project, uniform, pupil, classroom

University: seminar, tutor, lecture theatre

Both: canteen, exams, term*, coursework, assignment, timetable, library

Note *term* = British English; *semester* = American English

Extension: *Generating topic-related vocabulary*

Aims: *To explore derivatives and equivalent expressions related to the vocabulary in Exercise B.*

Preparation
- Ask students to bring their monolingual dictionaries. Bring some monolingual dictionaries yourself in case students do not have their own.

Procedure
- First deal with derivatives. Write *class* on the board. Elicit from the students other words that can be derived from *class*, e.g. *classroom, classmate.* Ask students to work in pairs, using dictionaries if necessary, to generate derivatives for the other words in the box in 1B. When they have finished, elicit suggestions and add them to the appropriate columns on the board.
  Suggested answers: *coursework*: *course; exams*: *examiner, examination, examine; lecture theatre*: *lecture, lecturer; library*: *librarian; tutorial, tutor*
- Now ask students to consider words in one area of education that may have an equivalent expression in another, e.g. *lecturer/teacher*. Students work in pairs to add any more expressions to the appropriate column.
  Suggested answers: *pupil, student; lecturer, teacher; assignment, homework; lecture theatre, classroom*

C ▸ Read out the instructions. Elicit a model from a student, or give an example yourself, e.g. (for daily student routine) *Most days we have a lecture at 9:00, followed by a tutorial at 11:30.*

▸ Give students a few minutes to think, before eliciting a few examples from the class. (You may like to introduce the notion of continuous assessment, in which final grades are awarded based on work done throughout the course, rather than solely on a final exam. This is particularly common for non-academic subjects, e.g. art or drama.)

Challenge

▸ With more confident groups, ask students to role-play the conversations the students might be having in the photographs. Encourage them to use the words in the box in their conversations.

IN THE EXAM

▸ Draw students' attention to the *IN THE EXAM* box. The information in this box is particularly relevant to students who are new to IELTS. Before reading through the box, tell your students to close their books and ask them what they know about the Listening exam in terms of length, number of sections, differences between the sections and the nature of Listening Section 1. Ask them to discuss in pairs and then check their answers by reading the box. Conduct feedback as a class.

2 Anticipating what you will hear

Aims: To encourage students to use visual and audio prompts to anticipate the language they will hear.

To introduce and practise form completion questions.

A ▸ On the board, draw two or three road signs which warn drivers of possible problems on the road ahead, e.g. *sharp bend to the left; beware falling rocks; road narrows,* or similar signs from the country where you are teaching. Ask students why these signs are a good idea. (Answer: *The driver is better prepared for what he or she may encounter.*) Tell students that good listening is a bit like good driving: you have to anticipate what is going to happen next. One of the key strategies for success in the Listening exam is to anticipate the situation and language you will hear. In this way you are better prepared for the listening tasks that follow.

▸ Direct students' attention to Exercise 2A. Explain that they can use the form on the page, in other words, what is printed on the exam paper in front of them, as a visual cue to help them anticipate what follows. Ask students to discuss the four bullet points with a partner.

▸ After a few minutes, get some feedback from the class, but do not confirm any suggestions just yet. Students will listen to a conversation related to the form in the next section.

ANSWER KEY

Suggested answers:

Situation: a student enrolling for a course

Speakers: student and secretary

Location: enrolment office, St. Vitus Academy

Language: *What is your name? Which course are you interested in?*

B ▸ The next thing that will help students anticipate the language they will hear is an audio cue. Before each section in the Listening module, they will hear a brief introduction, e.g. *'You will hear two students talking about their favourite subjects'.* They can use this introduction to confirm the ideas they already had from the visual cue. Hopefully, their initial ideas about the situation were correct. If not, they should be prepared to adjust their predictions accordingly. Tell students they will now hear the introduction to the form-completion task in 2A. Ask them to listen carefully and confirm or adjust their predictions accordingly.

▸ Play recording 2.1.

▸ Ask students if they predicted the situation correctly. If they didn't, ask them to take another look at the form. Were there any clues that they missed?

2.1 **LISTENING SCRIPT**

You will hear a student enrolling on a course.

C ▸ Tell students that like good drivers, good listeners never stop predicting what is coming next. Tell students they will now hear the first part of the dialogue. Ask them to listen closely and again confirm or adjust their predictions.

▸ Play recording 2.2.

2.2 LISTENING SCRIPT

(Sec = Secretary; St = Student)

Sec: Hello, come in.

St: Good morning. Are you the enrolment secretary?

Sec: I am.

St: I'm not too late to enrol on a course, am I?

Sec: No. We'll be enrolling new students till the end of the week.

St: Oh, thank goodness!

Sec: Have you done one of our courses before?

St: Oh no, this is my first time: the first step of my brilliant career.

Sec: Well, let's hope so. First though, we have to fill in a form.

St: Forms! They're so boring!

Sec: But necessary, I'm afraid.

D
- Read out the rubric for this exercise. The point of this exercise is to show students the extent to which answers can be predicted. For example, in Question 6 we may not be able to predict the course start date, but we do know that it is likely to be a figure. Give this as an example, then get students to work in pairs to make predictions about the type of information needed to complete Questions 1–5.
- Get feedback from the class.

ANSWER KEY

1 The student is female, so the answer will be a female name.

2 The surname will probably be spelled out.

3 Different countries give addresses in different ways. In most English-speaking countries the number of the house is given first, followed by the name of the street, e.g. *17 Church Road.*

4 The answer is likely to be the age of someone in their late teens to early twenties, as this is the age of the majority of students.

5 This will be the name of a field of study available at higher education level, e.g. *mathematics, physical education, business administration, history of art*, etc.

6 The date is likely to be in the form of a day and a month.

E
- Tell students they will now hear the entire dialogue. Ask them to listen and complete questions as they listen. Play recording 2.3.
- Ask students to check answers with a partner, or check them as a class.

ANSWER KEY

1 Sara; **2** Walker; **3** 19 Swan Street; **4** 18; **5** modern dance; **6** 5th September

Support

- As this is the first listening exercise, students may need some time to tune in. Therefore, you may like to play this recording twice or, with small groups, actually allow students to stop and start the recording as they wish. However, you should remind them that in the exam they will only hear the recording once – straight through!

2.3 LISTENING SCRIPT

(Sec = Secretary; St = Student)

Sec: Hello, come in.

St: Good morning. Are you the enrolment secretary?

Sec: I am.

St: I'm not too late to enrol on a course, am I?

Sec: No. We'll be enrolling new students till the end of the week.

St: Oh, thank goodness!

Sec: Have you done one of our courses before?

St: Oh no, this is my first time: the first step of my brilliant career.

Sec: Well, let's hope so. First though, we have to fill in a form.

St: Forms! They're so boring!

Sec: But necessary, I'm afraid. Now, what is your first name?

St: Sara, no 'h'.

Sec: Sorry?

St: S-a-r-a. Sara.

Sec: OK. Now your family name?

St: But I'm thinking about changing it to Simone, or maybe Sylvia.

Sec: Well, I think Sara's fine. So, what's your family name?

St: My family name is Walker.

Sec: W-a-l-k-e-r?

St: Yes.

Sec: OK. Now, where do you live?

St: Oh, yes, it's nineteen, one nine, Swan Street, London.

Sec: Swan?

St: Yes, like the bird.

Sec: OK, and your postcode?

St: N8 6BY.

Sec: N8 6BY ... And how old are you?

St: 18, well ... 19 next month.

Sec: So, 18. And what course would you like to do, Sara?

St: Well, I woke up this morning and said to myself, 'Now Sara, what are you going to do with your life? What course are you going to do?' My Dad thinks I should become a financial advisor and do people's accounts, because I'm pretty good with figures. Or perhaps I should be some kind of biologist, because that was my best subject at school. Or should I pursue my real love, modern dance?

Sec: And you chose modern dance.

St: Yes, I did. I want to make my living as a dancer.

Sec: Good for you. Now the course starts in the first week in September, which is ... let me see ... the fifth.

St: Oh, great. I can't wait.

Sec: Now, hold on! You have to be accepted onto the course first. You'll need to come in for an audition and interview.

St: Oh, no! I hate auditions.

Sec: I'm afraid if you want to be a dancer they're all part of the job!

St: Ah, well. When can they see me?

3 Following instructions carefully

Aim: To alert students to errors commonly made by IELTS candidates in the Listening exam.

- Tell students that IELTS candidates often lose marks in the Listening exam, not through having poor listening skills, but by failing to follow the instructions carefully. Ask them to read the rubric in the *Coursebook*. Then ask students to close their books and tell their partners what the three error types were, then open their books again and check.
- Ask students to look at the example exam question in the box. Tell students that the form has been completed, but it contains several errors. Ask students to identify the errors and error types by marking them A, B or C.
- When students have finished, tell them to check their answers against listening script 2.3 on page 118.
- Check answers as a class. Ask students to turn back to the form on page 17 and check for errors in their own answers. Did any other error types occur?

ANSWER KEY

1 No error

2 A
Note The rubric asks for no more than three words.

3 B
Note The address is 'Swan Street', not 'Road'.

4 No error

5, 6 C
Note These answers are in the wrong order.

4 Identifying keywords and paraphrase

Aims: To highlight the importance of identifying keywords and paraphrase before listening.

To raise awareness of and give practice in recognising distractors.

To introduce multiple-choice questions as a task type.

A

- Draw students' attention to the multiple-choice question in the box. Ask them what the different parts of this question are. Elicit the beginning of a sentence and the three choices. Say: *Take a look at this multiple-choice question. You are given the beginning of a sentence. This is known as the 'stem'. You are also given three alternative endings to complete the sentence. These are called 'options'. To answer the question, you have to choose the option you think fits the stem best.*
- Read out the rubric under the multiple-choice question, one question at a time. Allow students to discuss their answers briefly before eliciting their responses. Explain that, because the words used on the recording are rarely exactly the same as the words used in the questions, this process of identifying keyword(s) and paraphrase is key to IELTS success.

ANSWER KEY

The keyword in the stem is *career*. Paraphrases are: *job, occupation, do for a living, work as,* etc.

Paraphrase for options:

a accountant: financial advisor, consultant

b biologist: scientist

c performer: actor, singer, dancer, mime artist, circus artist

B ▸ Tell students that they may hear information that relates to all of the options. Warn your students that unless they listen very carefully they might choose an incorrect option as their answer. These incorrect options are known as *distractors*.

▸ Tell students that they are going to listen to part of the dialogue again and to make notes about what is said relating to each option. Ask them to identify which options are distractors and say why.

▸ Play recording 2.4.

▸ Ask students to check their ideas with their partner, then as a class. Point out to your students how keywords may be paraphrased (as in c) and how incorrect options may be directly referred to (as with B – *'some kind of biologist'*) or referred to using paraphrase (e.g. *a financial advisor*).

ANSWER KEY

A accountant – distractor
Note 'My Dad thinks I should become a financial advisor and do people's accounts because I'm pretty good with figures.' Note the use of paraphrase: 'a financial advisor'; 'do people's accounts'.

B biologist – distractor
Note 'Perhaps I should become some kind of biologist.' Note the use of 'should'.

C performer – the correct answer
Note 'I want to make my living as a dancer.' Note the use of paraphrase: 'make my living as' means the same as 'a career as' (see stem). Also 'a dancer' – in this context, is equivalent to 'a performer'.

Challenge

▸ Some students may have studied IELTS or a similar exam before and may already be familiar with multiple-choice questions and strategies for finding the answer. If this is the case with some or all of your class, test their knowledge by skipping Exercises A and B. Simply direct them to look at the multiple-choice question, play the recording and ask students to record their answers.

▸ Working in pairs or small groups, students should then compare their answers and explain how they came to choose that answer. Check answers as a class.

▸ Now teach the strategy by eliciting the meaning of the terms *stem, option, keyword, paraphrase* and *distractor*. Illustrate these ideas by referring to the multiple-choice question and, if necessary, working through Exercises A and B.

2.4 LISTENING SCRIPT

(Sec = Secretary; St = Student)

Sec: And what course would you like to do, Sara?

St: Well, I woke up this morning and said to myself, 'Now Sara, what are you going to do with your life? What course are you going to do?' My Dad thinks I should become a financial advisor and do people's accounts, because I'm pretty good with figures. Or perhaps I should be some kind of biologist, because that was my best subject at school. Or should I pursue my real love, modern dance?

Sec: And you chose modern dance.

St: Yes, I did. I want to make my living as a dancer.

Sec: Good for you.

5 Form completion

Aims: To consolidate skills introduced in this unit.
To provide further practice with form completion tasks.

for this task

▸ Ask students to close their books. Tell them they are about to do a form completion task. Remind them that in the exam, they are given time to read through the questions. Ask them to discuss in pairs what they would do in this time.

▸ Allow a few minutes for pair discussion, then elicit the following steps:

- Read the instructions carefully.
- Note the word limit.
- Look at the words on the page and try to anticipate the situation and language you will hear.
- For each answer, identify what type of answer the question requires and try to predict the answer.
- Identify keywords and consider paraphrase.
- Try saying unusual keywords and predicted answers to yourself.

▸ Refer students to the *for this task* box at the top of page 19. Ask them to note down any steps which they hadn't considered in their pair discussion.

▸ Now ask students to go through the pre-listening routine, using the form on the page. Give them 60 seconds or so, but remind them they will have less time in the exam.

express tip

- Read out the advice in the box while students follow in their books. Remind them that the wording on the recording is rarely exactly the same as the wording in the question. It is much more likely to be a synonym or paraphrase of the question. If the exact wording of the question is used on the recording, students should be careful that it isn't a distractor.

EXAM PRACTICE

Questions 1–7

- Play recording 2.5 through without stopping. This will show students just how long they have in the real exam.
- When the recording has finished, either check answers as a class, or go straight on to the next exercise, which relates to the same situation. If students wish, they can refer to the listening script on page 119, but they should wait until after listening to the complete section.

ANSWER KEY

1 Tufnell; **2** 7th July 1987; **3** 12, Castle Street; **4** OX4 2JP; **5** 72388; **6** Grade B; **7** History

2.5 LISTENING SCRIPT

(N = Nigel; R = Receptionist)

R: Hello! Hello, can I help you?

N: Yeah, I just wanted to … is this where you apply for courses?

R: Well, it depends what course you want to take.

N: Um … business administration.

R: Fine. Take a seat and we'll complete all the necessary forms.

N: Oh … thanks!

R: Now, what's your first name?

N: Nigel.

R: Nigel. OK. And last name?

N: Tufnell.

R: Sorry … ?

N: Tufnell. T-U-F-N-E, double L.

R: And when were you born?

N: When was I born? Er, 7th July 1987.

R: OK, and you're male. And you were born in the UK?

N: Yeah. British.

R: And your first language is English?

N: Er … yeah.

R: Address?

N: 12, Castle Street, Oxford.

R: Postcode?

N: Oh, no! I knew you were going to ask me that. I can never remember it! It's OX4 … OX4 … 2PG. No! OX4 … 2JP. That's it!

R: Sure?

N: Positive.

R: Your phone number?

N: Oh, that's easy! 01865 723, double eight.

R: Now, qualifications. Any 'A' levels?

N: Yeah. Three.

R: Subjects and grades, please.

N: Um … maths, B. Economics the same and then er, history …

R: And what did you get for that?

N: E.

R: Oh, dear. Well, at least you passed.

N: Yeah.

6 Multiple-choice questions with single answers

Aims: To consolidate skills learned in this unit.

To present further practice in answering multiple-choice questions.

for this task

- Ask students to close their books and tell their partners how they would tackle multiple-choice questions, both before listening and while listening.
- After a minute or so, ask students to open their books and check their ideas with the information in the *for this task* box. Again, at this point you could emphasise the procedure by eliciting ideas and writing them up on the board.

EXAM PRACTICE

Questions 8–10

- Now ask students to look at multiple-choice questions 8–10 and work through the preparation stages. Give them about 20 seconds for this, which is all they would get in the actual exam.
- Play recording 2.6 once only. Afterwards, ask students to discuss their answers in pairs, then conduct feedback as a class. Refer them to the listening script on page 119.

ANSWER KEY

8 A; **9** B; **10** A

2.6 LISTENING SCRIPT

(N = Nigel; R = Receptionist)

R: And it's business administration you want to do, isn't it?

N: Business administration. That's right, yes.

R: That's code, code … let me just check … yes, BA010. OK, the next course starts next semester. That's the fourth of October.

N: How long does it last?

R: A full academic year.

N: A year?

R: An academic year. That's around nine months.

N: Oh. OK. Good. Good.

R: Now all I need from you is a cheque covering the cost of the course, which is £2,500.

N: I didn't realise you'd need it today.

R: Well, you don't have to pay it all right now, but obviously you need to pay before you start the course and there are only a limited number of places, so …

N: Can I give you a deposit?

R: A 10% deposit would secure your place, yes.

N: Is there a bank round here?

R: There's a cashpoint. Go out of this building. Turn left. Go past the library, and the cashpoint is just by the main lecture theatre, opposite the canteen.

N: OK. I'll … I'll go and get the money.

R: You'd better hurry; the office is closing in five minutes.

N: OK, won't be long.

Extension: *Form completion practice*

Aim: *To provide further practice in completing forms, using authentic forms.*

Preparation

- You will need examples of forms in English, taken from authentic sources as far as possible. These forms could be application forms, contest entry forms cut from the newspaper, or forms printed off websites. If you do not have access to forms, you could make them yourself.
- Make one copy of each form for each student. If you want them to work with more than one form, make more copies.

Procedure

- Put the students into pairs. Give one copy of each form to each student, so that all students have their own form(s).
- Ask students to interview their partner and fill in the form with their partner's details.
- For further practice, ask students to change partners and repeat the activity.

Section aims:

- To teach students how to structure their writing by focusing on how to write an introduction and how to organise the main body of their report.
- To show students how to interpret and understand graphic information in order to pick out key information.
- To teach students to compare information within graphs and between two graphs.
- To introduce and practise Academic Writing Task 1.

1 Introduction

Aims: To provide background information on the higher education systems in the UK and Australia.

To prepare students for the charts which follow, illustrating relevant topics.

To generate discussion comparing students, own higher education systems to the UK and Australian systems, and to elicit related vocabulary to the topic.

- As a warm-up, ask students questions about their future study plans: *Who plans to study abroad? Where? Who is thinking about studying in this country? Where? What type of course do you want to follow?* etc.
- Elicit from the class what they already know about the higher education systems in the United Kingdom and Australia. Then, ask students to read the mini text and answer the three comprehension questions that follow. Check students' understanding of difficult vocabulary items such as: *postgraduate, pressure, commitments, challenging*. Finish off by asking students: *Why is a sandwich course called a 'sandwich course'? Do you think sandwich courses are a good idea or not?*
- Put students into pairs or small groups to discuss the similarities and differences between the higher education system in their own country, and in the UK and Australia. If students are too quiet in their discussion, prompt them to compare the 'student experience' in different countries as well as differences in degree classifications. Encourage them to bring in personal experience – have they or their friends/relatives been to university? Encourage students to give examples to illustrate their points.
- Monitor discussion and conduct brief feedback.

ANSWER KEY

A undergraduate and postgraduate; **B** a first; **C** balancing work and study; challenging job market; achieving a good degree

Challenge

- Ask students to read the text several times in order to memorise as much information as possible. Once they have done this, have them close their books and re-build the text by writing it down in their notebooks. Ask students to work in pairs or small groups. For feedback, ask one group to read out what they have written, then ask the class if any information is missing. If there is complete silence from the class, ask the student to read what they have written again, but this time have the class open their books and follow the text in the book as they listen. Now ask them again – *Was there anything missing*?

IN THE EXAM

- Ask students to read through the information in the box. Give students the chance to read through several times if necessary. Ask them to close their books and then ask the class comprehension questions such as: *How long should you spend on Task 1 in the exam? Why? How are you assessed? What's the minimum number of words required?*

2 Understanding visual information

Aim: To teach students how to use the title, key and axes labels to help them understand and interpret graphs.

- Ask students to study the graph and answer Questions A–E in pairs.
- When they have finished, check answers as a class, but do not get into a discussion about the information in the graph at this stage – the graph will be analysed further in the following sections.

ANSWER KEY

A Figure 1: Times Higher Education Supplement/Higher Education Statistics Agency

Figure 2: THES/Sodexho Lifestyle Survey 2004

B Blue shows the number of first class degrees achieved per year; red shows the number of upper-second degrees. The number refers to the total number of degrees awarded altogether.

C The vertical axis shows the number of students; the horizontal axis shows the year in which the results were recorded.

D The percentage of students who rated this problem as their number one worry.

E Total percentage is 88%. The remaining 12% of students might have cited a variety of other concerns that are statistically unimportant.

3 Writing the introduction

Aims: To teach students how to use the information in the key, axes and titles of charts to write their introduction to a Task 1 report.

To teach students how to write the introduction to a Task 1 report.

A

- Ask students to read through the introduction to Figure 1 in Section 2. Explain to students that their introduction for a Task 1 report need only be short, consisting of one or two sentences.
- Ask students to answer the true/false questions. After checking students' answers, emphasise to them that the introduction doesn't refer to the numbers within the charts – it refers to the information about the chart found in the title, axes and key.

ANSWER KEY

1 False; **2** True

Challenge

- To test students' ability to analyse and interpret key information within the graph, ask students to look at the information within Figure 1 and explain that three sets of numbers can be seen in this graph – firsts, upper seconds as well as the total degrees awarded. Ask them to match each of the following two descriptions to the three trends. Write these descriptions on the blackboard and ask students to work in pairs:

  1 A sharp rise between 1994–1995 and then steady rise until 2001. In the last three years numbers levelled off.

  2 A sharp rise between 1994 and 1995. From 1995 to 2000 numbers remained relatively unchanged. In the remaining years numbers again increased steadily.

(**Answer:** 1 upper seconds and total degrees awarded; 2 firsts)

- For students who are particularly good at analysing graphs, ask the following deeper comprehension questions: *Why do you think there was such a sharp rise in number of firsts and upper seconds awarded in 1995?* (because there was a sharp rise in the numbers of students graduating in 1994 compared to 1995) *Do the graphs show that students are more intelligent in 2003 compared to 1993? Why/Why not?* (Yes. This graph could be said to show this because while the number of students graduating has approximately doubled from 135,000 to 274,000, the number of students getting a first looks as though it has tripled from 10,000 to 30,000. Of course, an alternative explanation is that the students in 2003 were not more intelligent, but the exam papers had become easier or the marking has become more lenient!)

B

- Ask students to write the introduction to Figure 2 using the model example for Figure 1 as a guide. Remind students that they should NOT include any information from within the graph. They should just use the labels and title for information.

ANSWER KEY

Suggested answer:

The bar chart presents the main concerns facing higher education students in 2004, ranging from pressure to succeed due to financial cost to worries about achieving their desired degree classification.

4 Organising the main body text

Aims: To teach students what information to include in the main body of their report and how to organise it, with reference to cohesive devices.

To give students an opportunity to write the main body of a Task 1 report.

A

- Direct students' attention to Figure 2 and ask them what they think is the key information within the bar chart. Ask prompt questions such as: *How is the information organised within the chart? What numerical comparisons can you make between the different figures?*
- Ask students to read through the three sentences (a–c), which form the paragraph following the introduction describing Figure 2. Ask students to put them in the right order.

- Then ask students to identify the two synonyms for *concern* in the text (*worries and anxiety*) and explain that it is an important stylistic practice not to repeat the same word a number of times.
- Check that students have completed the task successfully, then have them look again at the underlined words in **a** and **c** and ask them what information these words refer to. After checking the answers, explain to students that using these cohesive devices links the information together and ensures that the writer is not repeating lengthy descriptors. Read out the sentences in order, substituting *this concern* with *achieving their desired degree classification* to show how clumsy this would sound.

ANSWER KEY

1 The correct order is: b, a, c
Synonyms: worry; anxiety

2 *This concern* refers to the students' biggest concern, i.e. worry about achieving the desired degree classification. *One* refers to students' biggest worry. *These figures* refer to 29%, 14% and 14%.

express tip

- Read out the advice in the box while students follow in their books. Impress upon them that they will be marked down in the exam if their report attempts to describe every bit of information in the graph – this usually indicates that the student is simply describing a graph without understanding or interpreting it. For example, looking at Figure 1, it would be a mistake to describe each and every change in student numbers, e.g. *From 1994 to 1995 there was a big increase in student numbers, then from 1995 to 1996 there was a smaller increase in student numbers. From 1996 to 1997 ...*, etc. This is obviously exhausting for writer and reader alike.

B
- Remind students that a Task 1 report must have a minimum of 150 words – we therefore need a second paragraph. Explain to students that while the first paragraph of the main body will include the main ideas, the second paragraph usually contains more in-depth analysis as well as surprising and/or less important information.
- Ask students to read through a model student's notes for the second paragraph and then answer the three comprehension questions (1–3).

ANSWER KEY

1 day-to-day financial worries, debt at graduation, pressure to succeed due to financial cost; **2** 26%; **3** balancing academic, social and work commitments, day-to-day financial worries

C
- Students should now be ready to write their own second paragraph based on the notes and answers to the comprehension questions. Have them do this in class, or for homework if time is limited. When students have finished, have them compare their answers with a partner.

ANSWER KEY

Suggested answer:

Not surprisingly, the question of money is a very big worry for students as it is mentioned in three separate categories (debt at graduation, day-to-day financial worries and pressure to succeed due to financial cost of university). Taken altogether, 26% of students rated these three categories as their number one concern. The joint third biggest concern for students is the day-to-day worries about money and the problem of balancing academic, work and social commitments, which 9% of students said was their biggest worry.

5 Comparing graphs

Aim: To provide practice understanding and writing reports comparing two graphs.

A
- Ask students to look quickly at the two graphs here, and answer any questions about them. Elicit the meaning of *sandwich course*, which they read about in the introduction to this unit.
- Put students into pairs and have them answer Questions 1 and 2. Monitor student discussions, guiding weaker students to the key parts of the charts and feeding in vocabulary and language of comparison as appropriate.
- Check answers by asking a pair of students to outline the main points. Follow this up by asking the class if they can add any more information.

Support

- If students are having difficulty interpreting the pie charts, then ask them the following true/false comprehension questions to direct them to key information within the charts. Alternatively, write the questions on the board: *There are the same number of full-time and sandwich students as part-time students.* (False: approx 1.5 million full-timers compared to 650,000 part-timers); *Post-graduate study is far more popular with part-timers than full-timers.* (True: 264,210 compared with 190,555); *Other undergraduates are the largest category of part-time students but the smallest category for full-time students.* (True); *There are approximately 10 times as many part-time students doing first degrees compared to full-timers.* (False: the opposite is true).

ANSWER KEY

1 The pie charts compare the breakdown in the number of full-time and sandwich students with the number of part-time students who studied in higher education in the UK in the academic year 2002–2003. Student numbers are broken down into three groups: first degree students, other undergraduates and postgraduates.

2 The key information is:

- There were many more full-time students than part-time (nearly double).
- The majority of full-time students were taking their first degree, whereas the minority of part-time students were taking their first degree.
- Other undergraduates were the minority of full-time students, for part-time students they were a much more significant group.

B

- Explain to students that they are going to complete the gaps to write the first paragraph of the main body of a report describing Figure 3. Ask students to first look at the vocabulary box and check understanding of any difficult words. Tell students to read through the paragraph a couple of times to fully understand it before trying to complete the gaps.
- Check answers by asking a student to read the paragraph aloud. Point out to students how the answer follows the broad principle of first describing general information from an overall perspective and then looking at the detail more specifically.
- Finally, ask students to finish the second paragraph in their own words. When they have finished, have them compare their paragraph with a partner's.

ANSWER KEY

1 from an overall perspective; **2** in terms of the figures; **3** the former; **4** more specifically; **5** in comparison to

ANSWER KEY

Suggest answer:

The majority of full-time and sandwich students were taking their first degree whereas for part-time students, it was the minority. As a mirror reflection, we can see that over 50% of part-timers were 'other undergraduates' while for full-timers, they represent roughly 10%. There was a similar picture for postgraduate part-time students who are more numerous both in absolute and relative terms than full-timers.

Extension: *Comparing and describing numbers*

Aim: *To teach students how to describe and compare numbers using the language of approximation.*

Procedure

- Explain to students that in written reports we don't normally describe statistics using exact numbers as this can be very boring and distracting for the reader. Instead we use approximation to round numbers up or down.
- Elicit different ways of describing the following numbers:
  134,575: *over 134,000, approximately/roughly/around/about 135,000;*
  134,575 compared to 396,530: *over triple, around 200% more, about three times as many, roughly 260,000 more, about a third as many, approximately one in three, far more, far less*
- Write the expressions used to describe approximations on the board. Now ask students to write up sentences using the numbers in the graphs: *There are just over 135,500 other undergraduates on full-time or sandwich courses. There are about three times more part-time undergraduates than full-time ones.*
- Ask students to go back to the charts on the previous page and write sentences about the information using approximations.

express tip

- Read out the advice in the box while students follow in their books. Explain to them that this is the way that people generally process information, i.e. they first need to see the big picture before looking at the detail and specifics. We need to do this in order to contextualise information.

C ▸ Put students into pairs again and have them look at the bar chart in order to answer Question 1. Then in pairs (or individually), get students to write an introductory paragraph. Check answers by getting one or two students to write their answers on the board or read them aloud. Highlight any errors and elicit corrections from the class.

▸ Ask students to read Question 3. Tell them that the four comparisons listed here represent some key information within the chart. Have them write four sentences, appropriately comparing the two institutions using information taken from the bar chart and table. Check answers as a class, or by getting students to write their answers on the board where you can do error correction work.

Support

▸ If students are having difficulty understanding what the graphs are about, then ask the class the following yes/no comprehension questions to direct their attention to particular aspects of the graphs: *Does the graph show the total number of students at each university?* (No, it shows total overseas students at each university.); *Does Middlesex have the least amount of overseas students in the UK?* (No, it has the 10th most overseas students in the UK – as a point of reference, there are over 100 universities in the UK); *Has Essex had a big increase in overseas student numbers?* (Yes, a 40% increase, the 10th highest in the UK); *Is the table organised according to total overseas student numbers?* (No, it's organised according to the percentage increase in overseas student numbers.)

Challenge

▸ For students who do well in this task, have them write up their sentences into two paragraphs – a suggested organisational structure might be to have Paragraph 1 as a comparison between graphs whilst having Paragraphs 2 and 3 as comparisons within each of the two charts.

ANSWER KEY

2 Suggested answer:

The bar chart shows the ten most popular education institutions in the UK for foreign students in the academic year 2002–2003. The accompanying table shows the top ten education institutions that have recorded the highest growth in overseas student numbers in the same year.

3 Suggested answers:

- Nottingham had the highest number of overseas students whilst Wolverhampton had the highest increase in overseas students.
- Wolverhampton had by far the biggest increase in overseas students in the UK, more than twice the increase than second place Salford.
- Middlesex had the lowest number of overseas students compared to other institutions listed, although it had the tenth highest in the UK overall.
- Nottingham had about one third more overseas students than Middlesex.

6 Academic Writing Task 1: Report

Aims: To give students the opportunity to consolidate the skills introduced in this unit.

To practise a complete Academic Writing Task 1 report.

To practise writing within a specified time limit.

for this task

▸ Ask students to read through the points in the box, which prepare them for the following exam practice question. Make sure students understand the information by checking any difficult vocabulary (*evidence, observations*) and ask comprehension questions to ensure they have absorbed the information.

EXAM PRACTICE

▸ The actual writing can be set as homework or done in class. As it is the first time students have done a Task 1 report, you could extend the time limit from 20 minutes to 30 minutes.

▸ When marking the work, look for:
- an introductory sentence which uses the title, axes and key to inform the reader of what is being described;
- a second sentence which says something very general about the chart;
- further statements about three major features of the graph, which are supported by statistical evidence.

▸ You should use the following criteria to assess the writing:
- task fulfilment: does the student describe the graph well?
- coherence and cohesion: is the report logically organised and linked together clearly?
- vocabulary and sentence structure

- When giving feedback to students on their essays, direct them to the model essay on page 106 of the *Coursebook*. Explain to students that the model essay is not a definitive answer – it is just one way of answering the question.

Model answer

The table illustrates the breakdown of scores for the IELTS Academic Paper in 2003. It shows separate scores for all four sections (Listening, Reading, Writing and Speaking), together with the overall score for students from four different language groups around the world.

From an overall perspective, Hindi speakers achieved the highest grades with an average score of 6.73 across all four sections. Moreover, they scored the highest of all four language groups in three of the four sections (Listening, Writing and Speaking).

Malayalam speakers scored the second highest scores overall, closely followed by Spanish and Russian speakers. Although Malayalam speakers did not do so well in the Reading, Speaking and Listening sections compared to Russian and Spanish speakers, there was a significant difference in their grades for the Writing section. These grades were high relative to Russian and Spanish candidates. Surprisingly, Spanish speakers, who achieved the second lowest results overall, achieved the highest results of all four language groups for the Reading section.

As a final point, it is interesting to note that the scores for each section show that all students on average scored the highest marks for the Speaking section and the lowest marks for the Reading section.

Technology

READING

Section aims:
- To teach students how to locate information quickly within the text, using a variety of skimming and scanning skills.
- To teach students how to quickly analyse tables and diagrams for the relevant exam task questions.
- To practise a number of task types: classification tasks; table completion; labelling a diagram.

1 Introduction

Aims: To introduce the topic of technology and to act as a lead-in to the reading passages.

To generate discussion on the theme of technology in order to elicit topic-related vocabulary.

- To start off the lesson, direct students' attention to the photograph of the couple watching the flat-screen TV and ask them questions such as: *What type of TV is this?* (Elicit *flat-screen, LCD TV* or *plasma TV.*) *Does anyone know the differences between a conventional TV, and a plasma or LCD TV?*
- Put students in pairs and have them answer the discussion questions. Deal with any difficult vocabulary (*technophobe*: someone who is afraid of technology and *technophile*: someone who loves technology). Monitor discussions and conduct feedback.

IN THE EXAM

- Draw students' attention to the *IN THE EXAM* box and read as they follow in their books. Ask questions related to the content of the box to check their comprehension.

2 Locating information in the text

Aims: To demonstrate how skimming a passage to understand the main ideas of each paragraph helps in locating information in the passage easily.

To prepare students for classification and table completion exam questions by teaching them how to navigate the rubric.

A
- Ask students to look at the title and photograph in the passage on the following page, and elicit what they think the passage is about (PDAs, or 'personal digital assistants').
- Have them read the five paragraph summaries in this section and explain that they need to match each one to one of the five paragraphs in the passage by skimming the passage. Have them complete the activity alone. In order to encourage skimming, rather than slow reading, set a time limit of one minute for this activity.

Support

- For students who are still hesitant with reading, take them through some standard steps:

  Step 1: Direct students to the title and photo of the text and ask them what they think the text is going to be about.

  Step 2: Ask students to read the first paragraph of the text on PDAs and have them choose the best summary prediction of what they think the text will be about?

  a History of PDAs; **b** Latest PDA features;
  c Comparison of PDAs on the market (answer: c)

  Step 3: Ask students to quickly skim the article to check their prediction – set a time limit of 30 seconds to 1 minute.

  Step 4: Ask questions to determine what type of writing the text is: *Is the text taken from a general interest newspaper article or a consumer magazine?* (From a consumer magazine – the writer talks about testing the two products in Paragraph 1.) *Does the writer want to provide information, or sell one or other of the two products?* (Provide information – the text is quite factually based.)

ANSWER KEY

2 e; **3** c; **4** a; **5** b

B
- Emphasise to students that the key to success in IELTS Reading questions is all about quickly locating the section where the answer is before contemplating what the answer is. Ask students to look at the table and decide in which paragraph they would expect to find the answer to each question. Students should work individually and compare answers at the end. Students should NOT read the text to do this exercise; they should instead draw on their knowledge of the text they gained in Exercise 2A.
- In order to complete the table completion task successfully, students first need to be able to understand how the table is structured, i.e. two

columns (positive and negative features) for each model and cross-referenced with the main ideas of each paragraph. Most students will pick this up – if they do not, this will need to be highlighted, i.e. Question 1 of the table is a positive feature of the ZV, which can be found in Paragraph 2 of the text about the pros and cons of the ZV.

- Direct students' attention to the five statements (a–e) and have them match them to the questions in the table. Explain that this task is designed to prepare them for the table completion question types that they will encounter in the exam. The aim is for them to be able to look at the table and decide what information they need to look for to complete the gaps.
- Model the first question as an example by asking the whole class. Elicit the correct answer and ask the student how they arrived at this answer. (The word *advantage* in the question corresponds to the column heading *Positive Features*, while *Palm Master* in the question shows that it is in the second row, i.e. Question 4.)
- Ask students to do the task individually and then compare answers with a partner.

ANSWER KEY

You would expect to find the answers in the following paragraphs:

a 3/4; **b** 3; **c** 2/4; **d** 2/3; **e** 2/4

The answers are: a 4; **b** 5; **c** 1; **d** 2; **e** 3

C

- Now ask students to complete the table alone, by reading the passage. Remind them that based on their answers to 2A and 2B, they know that Paragraphs 2, 3 and 5 are most likely to contain the answers to complete the gaps (1–5) in the table. This procedure helps students locate information quickly as they are not wasting time scanning the wrong part of the text.

ANSWER KEY

1 Wi-fi
Note 'The ZV comes with built-in Wi-Fi ...'

2 (digital) camera
'The drawbacks, however, are the 1.3 megapixel digital camera ...'

3 music and video
Note '... the restricted memory for storing music and video files...'

4 sharp clear screen
Note '... and a sharp clear screen to view data.'

5 docking station
Note '... the absence of a docking station ...'

express tip

- Read out the advice in the box while students follow in their books. Ask students to locate the footnotes in the table (*PDA*) and in the text (*WiFi* in Paragraph 2).

D

- Ask students to read through the classification exercise rubric in the box, and explain to them what they will need to do to answer this type of question. Students will need to decide which of the five statements refer to the ZV (A), which to the Palm Master (B), and which to both models (C).
- Explain to students that they will need to use scanning skills to answer this type of question efficiently. To this end, ask students to discuss Questions 1 and 2 before doing the classification task.
- Now ask students to do the classification exercise, then check the answers as a class.

ANSWER KEY

1 Because the product names appear too frequently throughout the passage.

2 The keywords in the text are: **1** Internet connection; **2** good value for money; **3** broad product range/variety of models; **4** optional extra; **5** PC

The answers are: 1 C; **2** C; **3** A; **4** A; **5** B

Extension: *Retaining and recalling information*

Aim: *To ask students to demonstrate their understanding and knowledge of the text in an enjoyable way.*

Procedure

- Put students into small groups of 3 or 4. Have them close their books and re-build the text by recalling as much information as possible. As a first step have them try to recall and note down the five paragraph summaries, then get them to take notes under these headings including as much detail as possible.
- It's a good idea to appoint one member of the group to take notes – this is a good opportunity to involve shy members of the group – they don't need to speak much in group work, but are the key person to record the group's discussions.
- Ask students to read out their answers to the class and compare their versions to the original text.

3 Linking visual information to the text

Aim: To introduce labelling a diagram as a task type.

A ▸ Explain the aims of this section to students and have them answer Question A in pairs or small groups. Encourage (even non-technically minded) students to guess one or two of the component parts and how it might work. This might prove difficult for some students, so use prompt questions, pointing to different parts of the diagram in the book such as: *What do you think this part is?* (pointing to Figure 5), *What does it look like?* (like a bottle or a gun), *What do you think it does? What are these lines coming from the end of it?*, etc. Avoid giving away any answers to the missing words in the labels at this stage.

B ▸ Ask students to answer Part B by skimming the text to find the paragraph section that describes the diagram. Set students a time limit of one or two minutes, depending on the level of your class.

ANSWER KEY

second half of Paragraph 3

Note 'The technology though, is tried and tested …'

C ▸ Before students answer the questions, explain that these diagram-labelling questions have a standard format which they should familiarise themselves with: the labels can be either completely blank (like Labels 2 and 3) or partially-completed (Label 1). Tell students that with partially-completed labels, they should use the words given to help them deduce the missing word and also to aid scanning.

▸ Ask students to answer the questions in pairs.

ANSWER KEY

1 clockwise; **2** Label 1–adjective, Label 2–noun

D ▸ As students have now completed the preparatory work, have them complete the labels in the diagram. Check answers as a class.

ANSWER KEY

1 phosphor-coated; **2** cathode element; **3** beams

4 Labelling a diagram

Aims: To provide further practice with diagram labelling tasks.

To give students the opportunity to consolidate the skills introduced so far in this unit.

for this task

▸ Ask students to read through the *for this task* box – this can be done with one student reading aloud while other students follow, or by having students read individually. Ask true/false comprehension questions to check understanding of the key points: *If there is no vocabulary box, you should take words directly from the text to complete the labels.* (true), *Grammatical connections between words on the label are not important.* (false), etc.

express tip

▸ Read the advice in the box while students follow in their books. Students should already know the information, but there is no harm exposing them again! Follow up by checking their understanding of skimming and scanning.

EXAM PRACTICE

Questions 1–5

▸ Ask students to skim the whole text several times (5–8 minutes) to understand the main ideas before allowing them to start the label completion task.

▸ Ask students to complete the labels alone. When they have finished, check the answers as a class. Students are now ready to complete the diagram.

Support

▸ With less confident students, before they start the exercise, ask them where the information is in the text to complete the diagram (top of column 2, page 33).

ANSWER KEY

1 front glass
Note '… on the front glass screen.'

2 RGB filters
Note '… sent through RGB filters ...'

3 liquid crystal cells
Note 'Each pixel contains liquid crystal cells …'

4 white fluorescent light
Note '… consisting of white fluorescent light which is shone ...'

5 micro-transistors
Note '… with tiny micro-transistors behind them ...'

5 Table completion

Aim: To provide further practice with table completion tasks.

for this task

- Ask students to read through the *for this task* box. Then ask them to close their books and recall as much information as possible in pairs. Emphasise to students that they should use words directly from the text to answer the questions and never invent their own words, or change word forms.

EXAM PRACTICE
Questions 6–12

- Ask students to complete the table answering Questions 6–12. They have done a lot of preparatory work, so you might want to set a time limit accordingly. (In the exam, students answer 40 questions in 60 minutes – just over one minute per question, plus 15–20 minutes to skim the three texts.) Go around the class, checking that students are referring to the correct part of the text for the answers.

ANSWER KEY

6 Biggest
Note 'Plasmas are taking the biggest slice of this market ...'

7 88%
Note '... they have an 88% slice of the market.'

8 Growth rate
Note '... relative to plasmas the growth rate is smaller.'

9 on the screen
Note '... that leaves static images on the screen ...'

10 prices
Note 'With bargain prices starting from ...'

11 screen(s)/screen size
Note 'Plasmas also beat LCDs in terms of maximum screen size ...'

12 easily wall-mounted
Note '... can also be easily wall-mounted.'

6 Classification

Aim: To introduce classification as a task type.

for this task

- Ask students to predict the content of the *for this task* box by asking them: *What are the key points you should keep in mind when answering classification type questions?* Elicit as much information as possible and then ask students to read through the box to confirm their ideas and reinforce the key points.

EXAM PRACTICE
Questions 13–18

- Ask students to do the exercise. Go around the class, ensuring that students are using the appropriate skills that they have learned, such as underling keywords.

ANSWER KEY

13 B (last paragraph); **14** B (Paragraph 3); **15** B (Paragraph 4); **16** C (Paragraph 4); **17** A (Paragraph 3); **18** B

Extension: *Recalling information*

Aim: *To demonstrate students' understanding and knowledge of the text in an enjoyable way.*

Procedure

- Ask students to close their books. Divide the class into three groups and assign a TV type to each group (Group 1: CRT; Group 2: LCD; Group 3: plasma). Tell students that they should note down as much information as they can remember about this TV type.
- As with the previous extension activity, appoint a note-taker for each group. Have them organise their notes under the following headings: *Economic picture/Consumer preferences; How the technology works; What's good about this TV; What isn't good about this TV; The future.*
- Ask the groups to read out their answers to the class.

Extension: *Topic sentence and paragraph summaries*

Aim: *To review topic sentences and paragraph summaries.*

Procedure

- After having skimmed the text, ask students to work in pairs to identify a topic sentence and paragraph summary for each paragraph.

Answers

Paragraph	Topic sentence	Paragraph summary
1	2nd sentence	The TV market is changing.
2	2nd sentence	Flat-screen TVs are big business.
3	1st sentence	Life expectancy of CRTs is limited but the technology is reliable.
4	N/A	Pros and cons of plasma, and cons of LCDs
5	1st sentence	Pros of LCDs
6	1st sentence	How plasma and LCD TVs work?
7	2nd sentence	The future of flat-screen TVs

PEAKING

Section aims:

- To provide additional practice of the Individual long turn and to encourage students to speak with confidence in this section of the exam.
- To provide language for, and practice of, giving a personal opinion.
- To provide language for comparing and contrasting information, and to give practice in identifying and answering questions requiring students to make comparisons.
- To practise Speaking Part 3: Two-way discussion.

1 Introduction

Aims: To introduce the topic of the unit.

To assess students' ability to answer Speaking exam, Part 3 questions.

- Use the discussion questions as a warm-up activity. Ask students to discuss the questions in pairs or small groups. When they have finished, ask for feedback from the class.
- Alternatively, put the students into pairs or groups of three to discuss the positive and negative aspects of technology. Allocate one of the discussion areas (work, family, transport or education) to each group. Ask the groups to prepare and give a mini-presentation to the rest of the class based on their findings.

IN THE EXAM

Draw students' attention to the *IN THE EXAM* box and ask them to read through the information individually. Point out that students will have to talk for 1–2 minutes in Part 2, so they should make sure they note down enough ideas to talk for this period of time. Part 3 will be covered in the second half of this unit.

2 Introducing the topic

Aims: To provide an opportunity for students to listen to candidates answering Part 2 questions.

To promote student confidence in the Individual long turn by providing them with language they can use.

A

- Explain to the class that they will hear four candidates answering a Part 2 question. They are responding to one of the Topics 1–3. Ask students to read the three topic questions.
- Play the three candidates' responses through once and ask students to identify which topic question is being answered.

ANSWER KEY

2 Describe an important email which you sent or received.

3.1 LISTENING SCRIPT

1

Candidate 1: Er, let me see: I'd like to talk about when I got an email from my father, you see I was travelling around the United States and I had asked him to send me some money …

2

Candidate 2: OK, I'm gonna tell you about an email I sent to my boss. Basically, I was unhappy about some of the things he was asking me to do. I mean, I had spent all morning photocopying documents, which really wasn't my job …

3

Candidate 3: Right, I'd like to tell you about the time I wrote an email to my landlord. The problem was that I had moved out of his apartment one month before, but he still hadn't returned my deposit. I decided that I would tell him …

4

Candidate 4: OK then, I want to talk about an email I got from an ex-boyfriend. Where shall I begin? It was about two years after we had split up and I was living in a different city and everything. Anyway, I got this email which said how he still loved me and that he wanted to try again, so I …

B

- Play the recording through again, this time asking students to listen for the language each student uses to introduce his or her talk. Ask them to write the language down in their notebooks. Ask students to discuss their answers in pairs before having a round-up with the whole class.

ANSWER KEY

Let me see, I'd like to talk about ...; OK, I'm going to tell you about ...; Right, I'd like to tell you about ...; OK then, I want to talk about ...

express tip

- Read out the information in the box while students follow in their books. Explain that the examiner will form a strong impression within the first few seconds of the candidate beginning to speak. It's therefore important that students begin to speak with a great deal of confidence – even if they don't feel it! Using a strong opener will make an excellent first impression on the examiner.

C
- Ask students to focus on the language in the box that they can use to introduce their Part 2 talk. You might like to play the recording a third time and focus on the intonation patterns of each speaker as they say the expression. Point out that a speaker's intonation often starts at a higher pitch when they start speaking.
- In pairs, ask students to practise introducing their talk, in response to Question 2 in 2A, by using the language in the box. Make sure they focus on the pronunciation of these ways to introduce the Individual long turn by drilling the intonation patterns. The sentences should start high and fall quickly. You could show this on the whiteboard by marking an arrow to show the rise and fall of the intonation. Get the students to either listen to the expressions on the recording again and repeat, or you could model the sentences yourself.

3 Individual long turn

Aim: To provide practice of the Individual long turn in the exam.

for this task

- Ask students to read through the information in the *for this task* box, then direct them to the *express tip* box next to it. Discuss them both together as a class.

express tip

- Read out the information in the box while students follow in their books. Impress upon students how little time one minute is and that they have to go with their first idea and make notes quickly. Explain that students should use the order of the points on the card as a frame to help them structure their talk coherently. Explain that writing full sentences is both impractical in terms of the time available and isn't useful, because they cannot be referred to easily when speaking. It is much easier and more useful to jot down keywords as a way to include impressive vocabulary and jog your memory when speaking.

EXAM PRACTICE

- Divide the class into pairs. Assign one student in each pair as Student A and the other as Student B.
- Student A will be a candidate and Student B an examiner. Have each pair practise Part 2 of the exam. Student A will have one minute to read Topic Card A and make notes before speaking.
- Encourage the examiner to assess the performance of the candidate by focusing on the criteria in the *for this task* box and *express tip* boxes, as well as the *IN THE EXAM* box on page 13.
- After a minute, have the examiner interrupt the candidate by asking one or two questions arising from the candidate's talk. When the first student has finished talking, have the examiner give feedback to the candidate based on the criteria described in the boxes.
- Then have each pair change roles and repeat the activity, with Student B answering the questions on Topic Card B.
- Go around the class and monitor the students closely during this activity. Try to include points students need to work on in a class discussion at the end.

4 Expressing opinion

Aims: To encourage students to express their own personal opinions.
To provide a wide range of language that students can use to introduce their opinions.

A
- Focus students' attention on the three questions related to technology. Explain that these are typical of the types of questions they will be asked in Part 3 of the exam. A successful candidate will need to answer these questions by expressing his or her opinion.
- Put students into pairs and have them discuss the questions. Stress that as well as saying what they think, they should also justify their opinion by giving a reason and/or evidence to support their opinion.
- When students have finished discussing Questions 1–3, elicit a few answers from the class.

express tip

- Read out the information in the box while students follow in their books. Students often use the word *because* to give a reason for something. Point out that when we speak naturally we don't often say this. We often don't use any conjunction at all – you could make a joke that students get more credit for not saying something!

B
- Ask students to read Questions 1 and 2. Choose a few students from the class and ask them to answer the questions with reference to their partner from the previous activity. They should focus on reasons their partner gave to support his or her opinions and the specific language he or she used to support those opinions.

C
- Explain to the class that they are now going to hear some students answering the same questions. Get students to focus on the language used to introduce their opinions by having them tick one of the boxes in the table. Note that all the language in the table is appropriate for students to use in the exam.
- Play recording 3.2 through once, while students tick the relevant boxes. Check answers as a class, then play again if necessary.

ANSWER KEY

In my view ...; I doubt ...; I believe ...; Personally, I think ...; I guess ...;
I'm not sure if ...; In my opinion ...; For me ...

3.2 LISTENING SCRIPT

1

Student 1: Yes, of course, it is now completely different – personally, I think email has made written communication much faster and cheaper than before. Firstly, you can write to someone on the other side of the world, and with one click, the information arrives seconds later, and secondly it costs you very little – just the connection to the Internet.

2

Student 2: Well, I'm not sure if this has improved the way students learn or not. I guess things are similar in that if students want to learn, they still have to go to class, read and make essays, etc, but I suppose on the other hand everything is now done on computer and is not written by hand. Learning is certainly more convenient though.

3

Student 3: It's a good question, but yes, I think it has. I can imagine a time very soon, when everyone will use videophones, it will be much more useful than only hearing a voice. If it is used a lot in business, you can have face-to-face meetings when you are very far apart. This will save a lot of time and money for travelling to meetings.

4

Student 4: I doubt that those people who can't use a computer will find office work easily. However, there will always be jobs where you don't need to use one, for example manual jobs where you use your hands to make things – like a construction worker or something like that.

5

Student 5: I believe things have changed dramatically. For example, it was only a few years ago that students would have to use books to find information, whereas today, the Internet is the first place students go. This must be a good thing, so yes, in my opinion, things have improved.

6

Student 6: In my view, it's not as easy to find a job if you cannot use a computer because you do not have the skills most companies require. I mean, you cannot communicate well if you cannot use email ... and another thing is perhaps you have to use a database.

7

Student 7: For me it's much easier to do research these days, you know, to find things. Before, it was more difficult to find things out. What I mean to say is, you had to read many journals to find the information you were searching for.

D ▸ Have the class ask and answer the same questions from 4A again. This time ask students to work with a different partner and focus on using some of the language from the table to introduce their opinion. Encourage students to use lots of new language and to try to expand their range of language by using previously unknown words and structures.

5 Comparing and contrasting information

Aim: To provide language for and practice with comparing and contrasting information.

A ▸ Explain that some of the questions in this part of the exam are designed to see whether students can compare and contrast information, i.e. talk about the similarities and differences between things.
▸ Use the example sentence to focus students' attention on specific language that can be used to make a comparison. Get the students to pick out the grammar of the sentence which makes the comparison (*more time-consuming than* ...).
▸ Elicit from the class further ways that we can make comparisons before moving onto the next exercise, which requires students to listen for language to describe the similarities and differences between things.

Answer key

in some ways ... is also more ... than ...

B ▸ Tell the class that they will hear four candidates justifying their opinion. Explain that their answers are written here, but some information is missing. They will need to listen and complete the sentences.
▸ Play recording 3.3 through once, pausing to allow students to complete the sentences. Ask students to compare their answers in pairs before giving feedback or allowing students to listen again.
▸ Have the students pick out language which is useful for comparing and contrasting information in each question. Have them compare in pairs before conducting class feedback.

Answer key

1 quicker; cheaper than; more time-consuming than
2 are similar in that; not as enjoyable as; one of the main differences is that
3 However; whereas; much cheaper
4 on the other hand

3.3 Listening script

1

Candidate 1: I think email has changed the way we communicate at work. Of course it's much quicker and cheaper than writing a letter. When we get to work, there can be lots of emails to respond to, so I suppose in some ways email is also more time-consuming than before.

2

Candidate 2: Going to a library and using the Internet are similar in that they are both great sources of information. Maybe I'm old-fashioned, but I prefer reading a book to an article on a computer screen. The Internet is fantastic, but it's not as enjoyable as going to a library. I suppose one of the main differences is that you can hold a book, you know, pick it up and turn the pages ...

3

Candidate 3: Video mobiles are great fun, I use mine all the time. However, they are expensive. My bill last month was $100, whereas normal phone calls are much cheaper.

4

Candidate 4: Training people to be computer literate is expensive in the short term; on the other hand, it will eventually benefit the national economy.

C ▸ Having highlighted the useful structures in 5B, now ask students to produce some sentences using the prompts in 5C. Monitor closely and encourage students to try to use a wide range of language, including any new words learnt in this unit.

- Conduct class feedback after the activity, getting students to record some of the sentences in their notebooks.

Support

- If your class has trouble getting started with this activity, do the first question as a class on the board. See the answer key for a suggested sentence.

ANSWER KEY

Suggested answers:

1 Communicating by email is much faster than communicating by letter.
2 Travelling by car is less romantic than travelling by train.
3 Traditional banking is less convenient than banking online.
4 Doing research with books is a lot slower than using the Internet.
5 Traditional shopping is more fun than shopping online.

6 Two-way discussion

Aims: To provide an opportunity to practise comparing and contrasting information and giving a personal opinion.

To practise Part 3 of the Speaking exam.

for this task

- Go through the *for this task* box. Ask students to cover the box and elicit information from the class. Ask: *What are some ways of justifying your opinion? How can you buy yourself some time if you can't think of an answer right away? What are some expressions you can use to compare and contrast two things?*

express tip

- Read out the information in the box while students follow in their books. This part of the exam is where students get the opportunity to really show off their knowledge of English by expressing their opinion. Stress that it is important to keep up-to-date with current affairs. You could use this tip as a way into discussing keeping up-to-date with current affairs and having an opinion on it. Encourage students to start forming an opinion about issues by asking themselves the questions in the box.

EXAM PRACTICE

- Put the students into pairs and assign the role of examiner to Student A and the role of candidate to Student B. Ask Student B to choose questions from the list at the bottom of the page.
- Explain that the questions can be asked in any order. Ask Student B to close their book so they cannot read the questions. Encourage Student B to give full answers; they will need to justify their opinion. Point out that some of the questions suggest that a comparison may be made in the answer; get students to listen carefully for that question and answer appropriately. Instruct the students playing the role of Student A to assess Student B in terms of the advice in the *for this task* box and *express tip*.
- When students have finished speaking, have them change roles and repeat the activity. At the end of the activity, conduct class feedback given by the examiners. This can be reassuring to students that other class members are experiencing similar difficulties. Avoid singling out individual performances at this stage.

IELTS Express Speaking DVD

- If you are using the *Speaking DVD* which accompanies *IELTS Express*, Section 3 – Part 2 of the DVD relates to the content of this unit. The DVD could be shown at the beginning of this lesson to give students a general idea of the content and format of the Part 2 of the IELTS Speaking module. Alternatively, it could be shown at the end of the lesson to provide a recap and demonstration of the material covered in this unit.

For more information on the *IELTS Express Speaking DVD* and how to integrate it into your lessons, see page 109.

The Workplace

LISTENING

Section aims:
- To raise awareness of 'signpost words' and the role they play in linking sections of a text, and to demonstrate how the use of signpost words can help students follow a description of a process, map or diagram.
- To introduce flowchart completion and label completion exam task types.
- To introduce and provide practice with Listening Section 2: Non-academic monologue.

1 Introduction

Aims: To introduce the topic of work and the workplace, and to elicit some of the vocabulary used later in the unit.

To give students the opportunity to compare and contrast different types of work and workplace.

To give students the opportunity to discuss personal preferences regarding work.

- Put students into pairs or small groups and give them a few minutes to discuss the three bullet points. While they are doing this, go round and help students with vocabulary.
- Conduct class feedback on students' ideas.
- At this stage you could highlight some of the useful vocabulary or expressions, e.g. *both/neither/in a similar way; while/whereas; personally/I'd prefer/indoors/outdoors/work from home/an open plan office*, etc.

Extension: *Fact Finding*

Aim: *To give students the opportunity to interact with fellow students in order to find out more about each other and to talk about their own work experience (if any).*

Preparation
- Before the lesson, prepare a list of questions about work experience. Photocopy enough lists so that there is one for each student. Questions could include:
- *Have you ever had a job?*
- *Where do/did you work? Describe the workplace.*
- *What do/did you have to do in this job?*
- *How long have you been doing/did you do this?*
- *Do/Did you enjoy it? Why? Why not? Is/Was it well-paid?*
- *Would you recommend this type of job?*
- *If you have never had a job, why not?*
- *Describe your ideal job.*

Leave a blank space for students to invent their own question.

Procedure
- This exercise lasts 10–15 minutes. You could do it following the introduction, or in place of it.
- Give out the list. Ask students to read through the questions. Point out that there is a space for them to write their own question. Have the students write a question in this space. Invite the students to go round 'interviewing' fellow students. With small groups they could interview everyone; with larger groups, divide the class into groups of three or four and get them to interview those students in their group.
- With small groups you could ask the class to report their findings on each student, or with larger classes, repeat their findings to another group.

IN THE EXAM
- Draw students' attention to the *IN THE EXAM* box and read as they follow in their books. Ask students to close their books, then test their knowledge.

2 Identifying signpost words

Aims: To present a range of signpost words, and to show how these signpost words help students to follow the description of a process.

To introduce flowchart completion as a task type.

A
- Ask students to look at the flowchart. Ask them what it shows (the procedure for booking meeting rooms). Ask how many stages there are (3). Tell students to quickly look over the flowchart to get a general idea of how the procedure works.
- Tell students they will now read on extract from a talk explaining the procedure. Students should read the

speech in the bubble and work with a partner, discussing what words could go in gaps a–d. Tell students not to worry about the blocked out words for the moment. After a few minutes, get some feedback from the class on their suggestions, but do not confirm any at this stage – they will listen for them in the next exercise.

- Tell students they will now hear the talk from which the extract was taken. Ask them to listen and check if their predictions for a–d were correct. If not, they should write in the correct word(s). Explain that some words have been removed from the recording. Tell them not to worry about these words just yet.
- Play recording 4.1 once only. After the recording, ask students to check answers with a partner and then as a class.

ANSWER KEY

a First; **b** And; **c** Next; **d** Finally

4.1 LISTENING SCRIPT

Hello. On behalf of myself and my colleagues at the buildings administration department, I'd like to welcome you all to the new company headquarters. Now, here in this brand new complex, we have a wealth of flexible facilities and space available for use. So, if you need a small meeting room for two people or a presentation suite for up to fifty, we at buildings admin. can help. However, we would ask that you don't just go into an empty room and start using it. We ask everybody to follow this simple room-booking procedure using the company intranet.

First, choose the sort of room you require and, most importantly, don't forget to tell us the time and ██████ you'll be needing it. And you might also like to let us know if you have any special requirements – conference calling facilities, for example. Coffee and other refreshments are always available. But if you need sandwiches, a buffet or a sit-down lunch, you need to contact the catering department. Next, fill in the booking form with your ████████████. This is an internal billing requirement, so please don't forget. Finally, you'll get confirmation of your room booking via ██████████. And that's it! Simple!

B ▸ Draw the following road signposts on the board and elicit from the class what each one tells the driver.

Ieltsville 25km

You are now entering Ieltsville
Please drive carefully

You are now leaving Ieltsville
Come back soon

- Explain that just like these road signposts, these signpost words (*first, and, next, finally,* etc) let us know where we are going, where we are, or where we have been. Tell students that like road signs, signpost words have different functions. Some functions are listed in the table.

Note Sequencing and listing:
Some students (and teachers!) may be confused about the difference between *listing* and *sequencing*. Listing (*firstly, secondly, thirdly, finally,* etc) identifies a position relative to *the complete list*. Sequencing (*before, after, next,* etc) identifies a position relative to *other items in the list*.

- Ask students to look again at the words for gaps a–d in Exercise 2A. Ask them to put each word into the appropriate space in the table. While they are doing this, draw a version of the table on the board. Elicit answers from the class, recording the correct answers on the board.
- Ask students to work in pairs to think of any other signpost words which have the same functions. After a few minutes or so, elicit answers from the class and complete the table. You may need to prompt students for some of the answers (e.g. *furthermore*: 3 syllables, first letter 'f'). Check pronunciation at this stage.
- For Question 3, ask students to follow the instructions in the *Coursebook*. Check ideas as a class.

ANSWER KEY

Listing: first, finally (suggested additional words: first of all, secondly, last)

Adding: and (suggested additional words: furthermore, also, as well, additionally)

Sequencing: next (suggested additional words: before, previously, while, meanwhile)

C ▸ Tell students they will now do a practice flowchart completion task. Direct their attention once again to the flowchart on page 38. Ask them to look at

Questions 1–3 and identify the answer type required for each gap (e.g. a time, place, reason, object, etc), the keywords and any synonyms or paraphrase they can think of. They should then try to predict the answers.

D
- Tell students they will now hear the complete talk. Ask them to listen and complete Questions 1–3 as they listen. Remind them to listen out for the signpost words, which will tell them when to move onto the next stage. Also remind them to listen for keywords, synonyms or paraphrase, as well as their predicted answers. Tell students to write their answers as they listen.
- Play recording 4.2.
- After the recording, get students to check their spelling.
- Check answers as a class, referring students to the listening script on page 121 if necessary.
- You may like to highlight a few items of vocabulary from the talk at this point, e.g. *a wealth of flexible facilities* (many adaptable spaces and/or services); *intranet* (similar to Internet but operates only within a company or organisation); *conference calling* (a system which allows people in different locations to 'attend' the same meeting via phone and/or video); *a buffet* (a selection of food laid out on a table; people help themselves and often eat standing up).

ANSWER KEY

1 date; **2** contact details; **3** email

4.2 LISTENING SCRIPT

Hello. On behalf of myself and my colleagues at the buildings administration department, I'd like to welcome you all to the new company headquarters. Now, here in this brand new complex, we have a wealth of flexible facilities and space available for use. So, if you need a small meeting room for two people or a presentation suite for up to fifty, we at buildings admin. can help. However, we would ask that you don't just go into an empty room and start using it. We ask everybody to follow this simple room-booking procedure using the company intranet.

First, choose the sort of room you require and, most importantly, don't forget to tell us the time and date you'll be needing it. And you might also like to let us know if you have any special requirements – conference calling facilities, for example. Coffee and other refreshments are always available. But if you need sandwiches, a buffet or a sit-down lunch, you need to contact the catering department. Next, fill in the booking form with your contact details. This is an internal billing requirement, so please don't forget. Finally, you'll get confirmation of your room booking via email. And that's it! Simple!

Extension activity: *Exploring signpost words*

Aim: *To raise general awareness of signpost words.*

Preparation

- On the board, draw the table below, or prepare it as a handout before the lesson.

Procedure

- Tell students that the three types of signpost words they have just met are not the only ones, there are others and other functions.
- Go through each function in the table, checking the meaning and asking for/supplying an example for each one.
- Now ask students to work in groups of three to come up with some more examples for each function (to make the task easier you could provide students with a jumbled list, which they then sort by function).
- When students have finished, elicit feedback and complete the table on the board. Check register and usage (see notes below), and pronunciation.
- If you haven't provided a hand-out, make sure students copy down the table from the board so they have a record.

Function	Examples
contrasting (showing difference)	but, however, although, though, yet, whereas, on the other hand*, while
cause and effect (what happens and why)	so, because, therefore, thus, consequently, as a result of this, as a consequence, this is caused by
exemplification (giving examples)	for example, for instance, such as, like
clarifying (making things clear)	in other words, what I mean is*, or to put it another way*
summarising (repeating the most important parts)	in summary, to summarise, so as we have seen, in a nutshell*

*These items have informal usage and students should not use them in formal or academic writing such as IELTS. They can be used in the Speaking section of the exam, and students should be able to recognise them.

Note Remind students that while these items may have similar functions, they are often used in different ways in different sentence positions (e.g. *although/however* and *but* are not completely interchangeable). For further information consult a good grammar book.

3 Following a description

Aims: To revise language describing the location of objects.

To introduce and give practice with label completion tasks.

A
- Ask students to look at the diagram. Stimulate interest by asking a few questions: *What is this?* (a sort of table on wheels) *Who might use something like this?* (somebody who has to move around in their job) Ask if they would find this object useful.
- Tell them they will hear a description of this object and will need to complete the labels. Ask students to work in pairs to answer Questions 1–4.
- Elicit answers from the class, but don't give away the answer to the labels just yet.

ANSWER KEY

1 The labels are arranged clockwise.

2 three words

3 1 is on (top of) the work surface; 2 is (directly) under/underneath/beneath the work surface. The work surface is on top of/on the adjustable stand./The adjustable stand is under the work surface; 3 is under/beneath/below the adjustable stand./The adjustable stand is on top of 3.

4 Student's own suggestions

express tip
- Read the information in the box while students follow in their books.

Challenge
- Some students may have studied IELTS (or a similar exam) before and may already be familiar with this kind of task and strategies for finding the answers. If this is the case with some or all of your class, test them by skipping Exercise A, directing them to the rubric for Questions 1–3, and play the recording as students answer.
- Working in pairs or small groups, students should then compare their answers and explain how they came to that answer. Check answers as a class.
- Quickly work through Exercise A and maybe play the recording again to ensure everyone understands this approach. Make sure you stress the idea of reference points and keywords.

Support
- You may wish to revise language such as *on top of, under(neath), above, below, to the left, on the left of,* etc. You could do this by placing an object (e.g. book or bag), in various positions. Elicit each position from the class. Students can then work in pairs, directing each other to move something from one place to another: e.g. *Put the bag under the chair. Now put it on top of the chair.*

B
- Play recording 4.3 and ask students to listen and answer Questions 1–3.
- When the recording has finished, ask students to check answers with a partner. Remind them to check spelling and the number of words used. Then check the answers as a class.
- Play the recording one more time as students listen and follow the completed diagram in their books.

ANSWER KEY

1 anti-slip; **2** pencil drawer; **3** five-star

4.3 LISTENING SCRIPT

Good morning, Ladies and Gentleman. Thanks for letting me come and talk to you today. My name's Bill Loman and, as most of you know, I represent Acme Office Supplies. It's our job to bring you the very best in office supplies and we do everything we can to make your working day just that little bit more pleasant.
Now, do you have to move around from one part of the building to another? Do you ever wish you could take your desk with you, along with your laptop and other essential bits and pieces? Would be great, wouldn't it? Well, now you can, with this – the Mobile Office! It's ideal for any situation where the user sits, stands or moves about.
Take a closer look at it. This is a ruggedly constructed,

highly adaptable unit. At the top is a flat work surface, where you can put your laptop computer and still have plenty of room for some files or documents. And you don't need to worry about your laptop accidentally getting knocked off because the laptop area is covered with a heavy-duty anti-slip rubber. There's even an optional laptop cable lock available. What's more, you can easily rotate the surface to cater for left- and right-handed users. Now, chances are you may have to carry around all sorts of other bits and pieces – pens, erasers, a stapler, a hole punch, etc. Well, the Mobile Office can help you out there, too, with a storage option under the work surface. There are a range of options available, depending on the model, but they include a single utility drawer, a double utility drawer or, as shown here, a pencil drawer.
Now, some of us are short, some of us are tall.
Sometimes we need to stand, others prefer to sit.
So the desktop work surface is mounted on a fully adjustable stand. This means you can vary the height of the surface between 30 and half and 42 and half inches. The stand is mounted on top of a five-star castor base, making the whole unit fully mobile. And if you are worried about the unit moving during use you can also opt for locking castors, keeping the whole thing as solid as a rock. The next thing I'd like you to see is this. Now, if you've got a lot of confidential documents, you can't just throw them away ...

4 Flowchart completion

Aims: To consolidate the skills learned in this unit by applying them to a flowchart completion task.

To give exam practice in Section 2: Non-academic monologue.

for this task

- Tell students they will now do an exam-practice flowchart completion task. Ask them to close their books and to tell their partners what they will do before, during and after they listen.
- Ask them to open their books and read the *for this task* box to check their ideas. You can also ask them to highlight the main parts in the *for this task* box.

express tip

- Read out the information in the box while students follow in their books. Elicit a few signpost words that were introduced on the previous page to see how many your class can remember.

EXAM PRACTICE

Questions 1–6

- Remind students of the word limit.
- Play recording 4.4. Do not give students any additional time to look at questions, nor should you give an explanation of the process.
- When the recording has finished, give students time to check their answers, spelling and word count.
- Check the answers as a class. Students might find it useful to hear the recording one more time, refering to the listening script on page 121 as they listen.

Support

- Some students may like to hear the recording twice. You can do this, but remind them that they will hear it only once in the exam.

ANSWER KEY

1 group; **2** presentation; **3** mental processes; **4** interview; **5** offered the position; **6** References

4.4 LISTENING SCRIPT

Part of my job as Human Resources Coordinator at the local Bob's Brushes factory is to recruit or select new members of staff. Today, I'd like to explain how our selection process works.
In the old days, if you went along for a job, you normally just got an interview. They asked you a few questions and that was it. But an interview does not give a true indication of a candidate's behaviour. For example, a candidate may say they are very good at doing presentations. But are they really? You need to see them in action. So, we like to put our candidates through what's called a recruitment process. This usually takes a whole day and consists of a number of stages. First of all, candidates are normally presented with a group exercise, because it's important to see if they can work well in a team. Then they are asked to deliver a brief presentation. Candidates are given advance notice of this so that they can prepare.
Next is a role-play. In this exercise, candidates are put into a difficult situation they might find themselves in if they are appointed to the position. Following this, candidates are set a series of psychometric tests, which assess their mental processes to see if they have the right skills for the job.
After all this, candidates attend an interview alone with a panel of three to four members of personnel from

different departments. At the end of the day, candidates are thanked and sent home with a bag of Bob's Brushes goodies.
Then comes the tricky part – we have to choose someone for the position. We analyse all the data and information we now have for each candidate. Sometimes none of the candidates are right for the job and we have to go through the whole process again. However, usually we find the right person and the successful applicant is offered the position. Unsuccessful candidates are also sent feedback on their performance. They may not have been successful this time, but we may need them in the future. The final stage is taking up references. Unless we discover from a referee that a candidate has lied on their CV, or has some kind of shady past, the candidate takes up their new position in a month or so. Job done. The post is filled. Now, any questions?

5 Label completion

Aims: To consolidate skills learned in this unit by applying them to a label completion task.
To give students further exam practice in Section 2: Non-academic monologue.

for this task

- Ask students to read through the *for this task* box, highlighting the key parts.
- Explain that students will now do a label completion task. Tell them that at first it may seem quite technical and complicated, but not to worry. They should find the starting point (Question 1) and focus on the labels and keywords rather than the machine. Remind them to follow the sequence of numbers and if they get lost they should listen out for reference points (labels that are already complete), which they can use as markers to find their place again.

EXAM PRACTICE
Questions 1–5

- Play recording 4.5.
- After the recording, give students a few seconds to check their answers, spelling and word count.
- Check answers as a class by playing the recording, but stopping just before each answer, elicit students' suggestions, then confirm by continuing the recording.
- Play the recording one more time while students listen and point to the relevant parts of the diagram. If necessary, refer students to the listening script for this section on page 122.

ANSWER KEY
1 input; **2** rotating; **3** (heated) metal; **4** guiding pins; **5** take up

4.5 LISTENING SCRIPT

Well, the next stop on our tour of Bob's Brushes is the production department and I thought I'd just show you this. Each and every brush which leaves the factory has our logo on the handle. In the old days, applying them used to be a manual task but these days it's all fully automated. It's called a 'logo machine' and at the moment it's set up for fixing logos onto the handles of 1½-inch paint brushes.
Let me tell you how it works. The paintbrush handles are placed into a large metal container just to the left of the conveyor belt there, called the 'input hopper'. This hopper vibrates, causing the handles to drop down a kind of funnel one by one onto a conveyor belt. This belt then carries the handles up into the main body of the machine. As they enter the machine, the handles move onto a rotating scroll. This rotating action realigns the handles and they are then carried along to the far end. At this point they are transferred to the handle wheel. This wheel carries the handles up in an anti-clockwise direction towards a metal roller, which is heated, and this is where the logo is applied.
Now, about these logos. The logos arrive from the suppliers on a long strip of special paper wound round a big drum. It works a bit like putting film into a camera. You see just to the right of the control station there? That's the 'logo strip payoff drum'. There're about 25,000 logos on that. From this, the logo strip is then fed round a series of what we call 'dancers', but what are technically referred to as 'guiding pins'. These are a very important part of the machine as it's crucial to maintain exactly the right tension in the logo strip. Anyway, the logo strip moves back towards the handle wheel and as it passes between the wheel and the roller, the logo is transferred to the handle. The waste strip, now empty of logos, carries on and is wound round the 'logo strip take up drum' on the far right there. And the brush handles? Well, after the logo is applied by the roller, they continue round and are unloaded through a special cylinder at the top left-hand side of the wheel, which blows them down a pipe into a waiting box at the back of the machine. Clever stuff: fast, simple, efficient.
OK, now have you ever wondered about how tooth brushes are made? Well, in the old days ...

WRITING

Section aims:
- To teach students how to decide between a thesis-led and an argument-led approach in writing.
- To teach students how to write an opening paragraph and conclusion for the thesis-led approach.
- To introduce different ways of supporting your ideas through reasons, statistics, scenarios or examples, as well as how to express disagreement of other people's views.
- To introduce and provide practice with Academic Writing Task 2.

1 Introduction

Aims: To generate discussion about the topic of employment.

To elicit topic-related vocabulary used later in the unit.

- Here is an optional activity before opening the book: get students to have a pyramid discussion on the most important ten jobs in modern society. A pyramid discussion works like this: first put students in pairs to decide on their top ten, then combine pairs into fours, then eights, and so on until the whole class has joined, amending their lists as necessary. At each stage when students disagree with one another, they need to justify their choice and re-negotiate the top ten. Put the agreed top ten on the board, practising pronunciation and checking meaning. Limit this activity to 15–20 minutes.
- Direct students to the photograph at the top of the page. Discuss the first question with the class.
- Ask students to discuss the following two questions in pairs. Then get students to share their ideas with the class, inviting comments from other pairs and getting students to justify their opinions.

IN THE EXAM

- Have the students read the information in the *IN THE EXAM* box one paragraph at a time. Ask them to close their books and ask concept questions to ensure students have understood: *What kind of questions ask for your opinion, not both sides of the argument?* (How far do you agree? To what extent ...?) *Where do you write your opinion in this type of essay?* (in the introduction, the main body and again in the conclusion) Get students to open their books again and highlight the important information in the box.

2 Recognising different approaches

Aims: To provide an example of the kind of question students will encounter in Task 2.

To teach students how to decide between the two styles of introduction: thesis-led approach (with the opinion clearly stated) and argument-led approach.

To provide a model introduction using a thesis-led approach.

- Read out the exam question and get students to consider their opinion. Quickly elicit subjects typically studied at school and get students to rate them on a scale of 'very useful' to 'not useful' for future careers.
- Put students in pairs and get them to read the two introductions and analyse them. Tell them to look at the purpose of each sentence and then to compare the similarities and differences of the two. Point out that both of them are good introductions but the writer is adopting a different approach in each. Draw attention to the fact that the first sentence is the same. The last sentence is what differentiates the thesis-led approach in Example A from the argument-led approach in B, where no opinion is stated.

3 Writing the opening paragraph

Aims: To introduce students to the structure of a thesis-led approach introduction.

To provide an opportunity to practise writing an introduction for this style of essay.

A
- Focus students' attention on the exam question, ensuring they understand what work-related stress is. Elicit some reasons why people are stressed at work (workload, relationships with colleagues, etc) and some symptoms of stress.
- Ask students to read through the three introductions and work through Questions 1–3 in pairs. Monitor and prompt where necessary.
- Go through the answers as a class, highlighting the importance of presenting an opinion explicitly and also the need to avoid copying the rubric word for word.
- Point out that examiners do not count words copied exactly from the rubric (unless there is no parallel

expression). Get students to highlight the paraphrasing in the introductions compared to the rubric.

Answer key

1 C; **2** A; **3** B

express tip

- Read out the information in the box while students follow in their books. Check students' understanding by asking students why they should not copy the question wording (it won't be counted) and what kind of essay requires a thesis statement (one where it asks to what extent/how far do you agree).

B

- Students should now be ready to write an introduction following the guidelines above. Tell them to use two or three sentences to write an introduction for the question in Section 2.

Support

- Before handing in their introductions for you to check, get students to pass them to a partner for peer checking. Did they avoid copying the question wording? Did they give their opinion? If the answer is not *Yes* to both of these, the student needs to try again before handing in their work to you.

4 Presenting and justifying your opinion

Aim: To provide students with four techniques for supporting an opinion in the main body of the essay: reasons, examples, scenarios and statistics, and to give guided practice in each.

A

- Remind students of the essay question on work-related stress in Section 3. On the board write the four techniques for supporting ideas: *reason, example, scenario, statistics*. Don't worry about meaning at this stage.
- Read **a** as a class and ask students which technique of the four it uses. Elicit the word that tells you it is a reason. Ask students to underline or highlight the word *because*.
- Let students do **b**, **c**, **d** and then check as a class, eliciting the clues in each.
- Go back and ensure students understand the meaning of the four techniques on the board.

Challenge

- If your class is strong, elicit other language for these four techniques.

Giving reasons: *since; as; because of; due to; this is because …; one reason for this is …*
Giving examples: *for example; such as; one example is …*
Scenarios: *imagine …; let's say …; just suppose that …*
Statistics: *research shows; a recent survey found; the majority of people …*

Answer key

a 3; **b** 1; **c** 4; **d** 2

B

- Ask students to read the instructions to the task, then look at the three prompts. Have them decide which of the three methods of justifying an opinion they would use with each sentence.
- Check and elicit what clues helped them decide.
- Ask students to work in pairs to complete each prompt.
- For feedback, get several students to write their answers on the board for each and check/compare. Ask students to decide if they have followed the correct technique and if the language is accurate.

Answer key

'Work-related stress is a problem because …' – reason
'If work-related stress is not dealt with …' – scenario
'The psychological effects of stress can be extreme. For instance …' – example

C

- This exercise gives students more freedom to practise the four techniques from the previous exercises. Let students choose which technique they use for each, but try to get them to experiment with two or three of the techniques. Get them to share answers.

Support

- Read Statement 1 and give students thirty seconds in silence to decide if they agree or disagree. Get students to 'vote' for agree or disagree and make two groups accordingly. In these groups students should discuss the statement and their reactions and opinions to it.
- Set each group the task of giving a reason for agreeing/disagreeing with the statement. After a few minutes, get them to share with the whole class. Then ask them to provide an example for the same

statement. Then get them to give a scenario and statistics if possible. (this one is trickier as they may not know any statistics off the top of their heads)

- Now divide the class into 3 groups (or 6 if a large group) and allocate one of the remaining statements 2, 3 and 4 to each group. Get students to follow the same procedure as above. Monitor and prompt. Get class feedback on all four statements.
- Finally, get students to write up their opinions for two of the four statements as consolidation. This could be given for homework if you are running out of time.

5 Expressing disagreement

Aims: To show students how to disagree with a point of view.

To present and practise the language of refuting an argument.

Support

- If you have a weak class, you could skip this section entirely for the moment, as it may confuse them.

A

- Read the instructions as a class, reminding them of the thesis-led approach where only one point of view is necessary. Point out that refuting the opposite opinion is optional, but something that good writers tend to do.
- Go through the language in the box, eliciting which words/phrases are used to refute an argument (e.g. *however, although*) and get students to highlight or circle these.
- Read the instructions for the exercise and do Question 1 together. Ensure students do not put *but* in the middle of the sentence as *although* serves the same purpose.
- Let students work through the sentences in pairs. Go around monitoring the students and prompt where necessary.

Support

- If this activity is too difficult for your students, use the suggested answers below to make a list of endings. The beginnings of Sentences 1–6 in the book could be written on one colour of paper and the endings on another for students to match beginnings with an appropriate ending.
- Alternatively, just write the endings on the board/OHT and mix them up.

ANSWER KEY

Suggested answers:

1 ... many employees do not work hard because their performance is not monitored.

2 ... there are others who would perform better if their work was rewarded financially according to results.

3 ... it doesn't mean that it is not beneficial to the education of children.

4 ... I would like to point out that people are still required to check that everything is working properly.

5 ... there are many people who work for little or no money at all.

6 ... targets and expectations of productivity have increased, so people are still putting in the same number of hours.

B

- First elicit what *performing arts* are, and ask students' opinions about whether they are a waste of time.
- Whatever their opinion, students need to come up with a reason for the opposite point of view, i.e. that studying performing arts is not a waste of time. Invite ideas as a whole class and write them on the board. Get students to choose the best argument and then make sentences using the language in the box. Have them try to say the same thing in two or three different ways to experiment with the language.
- Look at the second question and follow the same procedure as above. Make sure students can articulate their basic response before attempting to use the language. Get students to use different language this time to ensure they expand their repertoire.

ANSWER KEY

Suggested answers:

a Although it may be true that people do not use drama skills directly in their work lives, education is more than about preparing children for work.

b The argument that women should only be home-makers is an old-fashioned one and is a discriminatory remark against 50 per cent of the adult population.

6 Writing the conclusion

Aims: To provide students with a model conclusion in the thesis-led approach.

To teach students to use paraphrase in the conclusion to re-state the opinion they presented in the introduction.

A

- Remind students of the introduction of the essay about work-related stress from Section 3 on page 43. Elicit from the class what they think the writer planned to cover in the main body.
- Ask students what the purpose of a conclusion is (*to summarise what has already been said, but not to introduce new material*).
- Ask students if they think the words/phrases in the conclusion should be exactly the same as in the introduction or just the same ideas. (*same ideas but using paraphrasing*)
- Ask students to read the two conclusions and answer the questions in pairs. Monitor and remind them to compare the conclusion to the introduction. Elicit the paraphrasing and write it up on the board, along with the phrase from the introduction in one column and the phrase meaning more or less the same from the conclusion in the right column.

ANSWER KEY

Conclusion 1 rephrases the introduction and summarises the main points.

Conclusion 2 contains new information and doesn't paraphrase the introduction.

express tip

- Read out the information in the box while students follow in their books. Get them to cover the tip, and elicit the two key principles mentioned here, i.e. don't include any new information and use paraphrase.

B

- Get students to look again at their own introduction from Exercise 3B. First get them to experiment with paraphrasing the ideas before putting them together in a conclusion of about two or three sentences.
- Before students hand in their work for you to check, get students to swap work and check that their partner has not included new information, but has used paraphrase where possible.

Extension: *Writing the main body*

Aim: *To give students practice in writing the whole essay, planned above.*

Preparation

- You will need an overhead projector and enough transparencies for each student or pair of students.

Procedure

- Get students to look over their introduction, conclusion and ideas for supporting and refuting an argument.
- In pairs or individually, get students to write two or three paragraphs for the main body of this essay on an overhead transparency to share with the rest of the class. This should not take long as they have already planned the bulk of it.

7 Academic Writing Task 2: Essay

Aims: To give students the opportunity to consolidate the skills introduced in this unit.

To practise a complete Task 2 writing task.

To practise writing within a specified time limit.

for this task

- Ask students to read through the points in the box preparing them for the following exam practice question. Before they start the task, elicit the main points in the *for this task* box as consolidation.

EXAM PRACTICE

- Read through the question together. Elicit some high-level positions in companies. Discuss whether this situation is in fact true in their country as there may be a cultural gap. Explain *allocate* and discuss the idea of positive discrimination in the workplace.
- Either set the exam question for homework or do it in class. Depending on the level of the class, you could extend the time limit from 40 to 60 minutes, as this is the first time students will have seen and done a complete Task 2 Writing task.
- When giving feedback to students on their essays, direct them to the model essay on page 107 of the *Coursebook*. Explain to students that the model essay is not a definitive answer; it is just one way of answering the question.

Support

- Brainstorm ideas for the main body of the essay in groups or as a class to give students some ideas. Then set the task either for homework or in class as a timed open-book essay.

express tip

- Read the information in the box while students follow in their books.

Model answer

In many countries these days, females make up over 50 per cent of the workforce, and increasingly highly-skilled women are taking managerial positions. However, it is still a fact that high positions such as CEO jobs are still dominated by men. Although this is not desirable, I do not personally believe that imposed quotas are the solution.

Firstly, I believe companies have a right to choose the best person for the job, whatever their gender, in order to contribute to the success of their business. Forcing companies to hire, promote and appoint women could negatively affect business in the short term and even in the long term.

Secondly, to my mind the solution to this problem should be solved outside the workplace. Girls need to be encouraged to take more male-dominated subjects at school and later at university, and to aspire to do well in their careers. Girls and boys also need to be taught equality from an early age. This education can take place in schools, career programmes and in the home.

To those who argue that quotas are a good way to initiate this change, I would like to point out that artificially imposing rules has not always had the desired effect. When governments required males and females to receive the same pay for the same job, employers simply changed job titles to ensure that women were still paid less than men. It is my belief that employers will simply try to find loopholes to get around any such law.

In summary, I do not believe that forcing companies to allocate jobs to women is the best way to address this imbalance. Rather, it is a question of education and of changing mindsets so that those who deserve to be at the top will earn it and be appropriately appointed.

Climate and the Environment

READING

Section aims:
- To encourage a logical thought process of analysing meaning in order to answer Yes/No/Not given questions.
- To introduce the key skill of identifying paraphrased information.
- To practise the exam practice task types: Yes/No/Not given; sentence completion questions.

1 Introduction

Aims: To initiate discussion on the topic of the climate and the environment.

To generate discussion relevant to the topic of the reading texts later in the unit.

A
- Use this exercise as a way of leading into the reading lesson by having students discuss how environmentally friendly they are. You could begin by writing *Environmental issues* on the board and eliciting specific issues from the group, e.g. *global warming, car emissions, cutting down of rainforests, the developing world's energy requirement, industrial pollution, endangered animal species, disposing of waste in land-fill sites, the increasing numbers of fights, acid rain,* etc.
- Have the class read through the statements in the *green-o-meter* quiz. After you have checked any unknown vocabulary, ask students to agree or disagree with the statements by ticking the boxes.

B
- Ask students to discuss their answers with a partner. As they compare their answers, they should justify their opinion. After pairs have had a chance to talk, select students to tell the class which of the actions they do and which of the actions are more important than others.

Answer key

2 attitude: b, c, e, f, j; **action:** a, d, g, h, i

C
- Extend class feedback by having an open class discussion about other ways that individuals can help to save the environment – you could go on to compile a list of the top ten. Include in your discussion writing letters to politicians as this will lead into the extension activity below.

Extension: *Mini-presentations: Saving the environment*

Aim: *To give students further practice in describing a situation, explaining a problem and presenting an argument.*

Preparation

- On the board, write simple structures students can use to structure their presentation:

Greeting: Good morning, my name's ...

Introduce topic: Today I'm going to tell you a little about ...

Outline: First I'm going to talk about X, then move on to talk about Y, and finally have a look at Z.

Main body of talk: First I'd like to talk about ...; OK, now I'd like to move on and look at ...; Finally, I'm going to explain a little about ...

Conclusion: So, as we have seen ...; Now, I wonder if anyone has any questions?; OK. Thanks for listening, I hope you found it interesting.

Procedure

- Refer students back to the list of environmental issues facing the world they came up with at the beginning of the unit. Put the students into groups and allocate an issue to each group.
- Have each group look at the following three questions in order to explore the issue more fully: *What is the problem exactly? What are the causes of the problem? What are some possible solutions?* After they have discussed these together, ask them to explain to other members of the class, either in a mini-presentation format (using some of the language above), or by changing groups and explaining to each other what they discussed.

Extension: *Writing to a politician*

Aim: *To further reinforce the topic of this unit by giving students the opportunity to write a formal letter.*
(Although the academic module within the IELTS exam does not require students to write a letter, this could be an engaging and motivating task where students are writing to a real person.)

Preparation

- As well as having some input for students on how to structure a formal letter, you need to have an address of someone in power to write to. If you have access to the Internet, have the students research someone in government (in your country or another) who they can write to.

Presentation

- Explain to the class that they are going to write a letter or email to a politician – this could be a local politician or a country leader like the US president.
- Ask the class to discuss who they want to write to and about what issue. Provide input as to how to structure a formal letter requesting action.
- After students have written their letters, make sure that the letters are posted or emails are sent, as this will be very motivating for students.

IN THE EXAM

Draw students' attention to the *IN THE EXAM* box and read as they follow in their books. Explain briefly about the two question types covered in this unit. Do not go into too much detail at this stage as they are explored more fully in the *for this task* boxes later in the unit.

2 Analysing meaning

Aims: To introduce Yes/No/Not given questions as a task type.
To introduce a logical thought process in order to answer Yes/No/Not given questions.

A
- Focus students' attention on the title of the passage and have them predict some of the possible ideas contained in the text. Then get them to skim the text to identify the main idea.
- Have them compare their ideas in pairs or groups of three before conducting class feedback.

ANSWER KEY

The main idea of the passage is that global warming has secondary effects other than just rising sea levels.

B
- Explain that often when students have to deal with an IELTS reading text, it will probably be necessary to read only some sections very carefully and that in this case the second paragraph contains the information necessary to answer the following three questions.
- Go through the rubric at the beginning of Section B, emphasising that in order to answer 'No', the idea expressed in the text must be the opposite of the message in the text. Explain that an idea may be alluded to, but may not be the same idea as the one expressed in the text – in this case the answer will be 'Not given'.
- Ask students to read through Paragraph 2 very carefully and answer Questions 1–3. You could do the first one with them, making sure students follow the thought process laid out in the three bullet points.
- After they have finished, ask students to compare their answers and their thought processes before providing feedback to the class.

ANSWER KEY

1 Not Given; **2** Yes; **3** No

C
- Students should now answer Questions 1–3 in Section C following similar thought processes. Again, have them compare their answers and say how they arrived at that answer by explaining their thought processes to their partner. Conduct class feedback.

ANSWER KEY

1 Not given; **2** Yes; **3** Yes

Support

- This is quite a challenging text. One way to make it easier for students to get a grasp of what is being described is to draw a diagram of the ocean cycle on the board. Alternatively, get students to draw it themselves in pairs.

3 Identifying paraphrase

Aim: To reinforce the key skill of identifying paraphrased information.

A ▸ Ask students to focus on the title of the passage and picture, and predict what environmental problem the text discusses (the burning of coal as a source of energy).

▸ Refer students back to points **f** and **j** in the survey in the introduction (i.e. *Technology can solve all our environmental problems* and *Governments should use cleaner sources of energy*) and ask students to recall some of the arguments that were put forward. They should consider whether some of their arguments might be included in this text.

B ▸ Ask students to skim the text to get the main idea of the passage. They should compare their ideas in pairs or groups of three before having class feedback.

C ▸ Remind the class that many of the questions and answers in the IELTS exam are based around synonyms and paraphrased information. Ask them to match the words in the left column with their synonyms on the right, then compare their answers with a partner.

▸ Check answers as a class. The synonyms all come from the questions which follow, so refer back to these pairs of words to illustrate how identifying these synonyms is a key skill.

ANSWER KEY

1 e; **2** d; **3** b; **4** a; **5** c

4 Yes/No/Not given

Aim: To provide practice of Yes/No/Not given questions.

for this task

▸ Ask students to read the information in the box. Answer any questions they may have. Emphasise, in particular, the difference between *No* and *Not given* answers as these often confuse students.

EXAM PRACTICE

Questions 1–7

▸ Ask students to answer Questions 1–7 alone before comparing answers with a partner.

▸ In your class feedback, ask students to explain the thought processes that led them to arrive at the answers in the same way that they did in Section 2B on page 53. Refer to these thought processes when making clear the distinction between 'Yes', 'No' and 'Not given'. Make sure you focus on any synonyms or paraphrase which helped them identify the information in the text.

ANSWER KEY

1 Yes
Note 'In poor countries … air pollution is one of the leading but preventable causes of death. It affects the rich world, too …'

2 No
Note 'blame [for encouraging dirty technologies] can be laid at the door of the rich and poor countries alike …'

3 Yes
Note 'As the poor world grows richer in the coming decades and builds thousands of power plants, many of these people will get electricity.'

4 Not given

5 No
Note 'This should not be regarded as charity, but rather as a form of insurance against global warming.'

6 No
Note '... one good idea is for governments to impose a tax based on carbon emissions … with the revenues raised returned as reductions in, say, labour taxes.'

7 Not given

express tip

▸ Read the information in the box while students follow in their books. It can be especially tempting to do this if students are running short of time. Explain that the more obvious answer may not be the answer stated in the text, therefore students should avoid the temptation to simply answer whatever they think *should be* the correct answer.

5 Sentence completion

Aim: To introduce and provide practice of sentence completion as a task type.

for this task

▸ Ask students to read the information in the *for this task* box. Make sure they understand the two different types of sentence completion task. You should explain that both types of tasks require the students to locate the relevant part of the text quickly and then read the relevant information again carefully.

EXAM PRACTICE

Questions 8–12

- Ask students to complete the sentences with words taken from the text. Point out that they should use a maximum of three words, which they should take directly from the text, i.e. using the same words in the same form.
- After students have completed the sentences, ask them to compare their answers in pairs. Encourage them to show their partner the information in the text which led them to their answer. Finally, check answers as a class.

ANSWER KEY

8 (fossil) fuels
Note '… burning fossil fuels is having a massive effect on climate change.'

9 causes of death
Note '… air pollution is one of the leading but preventable causes of death.'

10 impose a tax
Note '... one good idea is for governments to impose a tax based on carbon emissions.'

11 low-carbon technologies
Note '… they would provide a much-needed boost to the development of low-carbon technologies.'

12 fuel-cell cars
Note 'Within a few years, nearly every big car manufacturer plans to have fuel-cell cars on the road.'

express tip

- Read the information in the box while students follow in their books. Explain that the order of the questions can act as a map to help the students find their way around the text and therefore, locate the information more easily. If a certain answer is contained in Paragraph 4, you know that the next answer cannot be in Paragraphs 1, 2 or 3.

EXAM PRACTICE

Questions 13–16

- Ask students to match the sentence stems with their endings.
- After students have completed the sentences, get them to compare their answers in pairs. Encourage them to show their partner the information in the text which led them to the answer. Check the answers as a class.

ANSWER KEY

13 E
Note '… governments must stop subsidies and exemptions that actually encourage the consumption of fossil fuels.'

14 F
Note 'As the poor world grows richer in the coming decades and builds thousands of power plants … However, many of these plants will burn coal in a dirty way.'

15 H
Note 'The final and most crucial step is to start pricing energy properly … That would make it absolutely clear that the time has come to stop burning dirty fuels such as coal.'

16 A
Note '… "sequestration", an innovative way of using fossil fuels without releasing carbon into the air.'

PEAKING

Section aims:

- To focus on the functions of the speaking card in Part 2 of the Speaking exam, namely describing and explaining.
- To enable students to recognise and respond to Part 3 questions which require a speculative answer.
- To enable students to communicate their ideas more clearly through using a number of signpost words and expressions.
- To practise Speaking Part 2: Individual long turn and Part 3: Two-way discussion.

1 Introduction

Aims: To generate interest in the topic of climate change.

To provide students with an opportunity to speculate about a subject.

- Write *climate change* on the board and elicit from the class what they know about the topic.
- Divide the class into pairs or small groups to discuss the two questions among themselves, before opening the discussion to the whole group.

Support

- Elicit a list of keywords from students to use in their discussion. The list should include: *global warming, rising temperatures, rising sea levels, floods, melting ice caps, agricultural patterns, crops, insects, violent storms, hurricanes, animal species, extinction* and any others that you think will help students discuss some of the issues. If necessary, encourage them to look up any meanings they are unsure of in their dictionary.

IN THE EXAM

- Draw students' attention to the *IN THE EXAM* box. Explain how the notes that the student makes are essential in structuring their talk. Explain that candidates should talk about something personal and that in order to be able to communicate clearly, they should 'paint a picture' with their words. To help them do this, they can close their eyes and visualise what they are going to talk about before they begin to make their notes.

2 Describing and explaining

Aims: To show how the speaking card tests two speaking functions: describing and explaining.

To provide language for introducing the explanation part of the card.

To provide an opportunity to practise Part 2 of the Speaking exam.

A

- Students have come across this part of the exam in Units 1 and 3. Remind students that they have to speak for 1–2 minutes about the topic on the card. Use the example card to illustrate that they have to both describe and explain something about the situation.
- Focus students' attention on the card and go through the questions. At this point you could ask students to think about what experience they would talk about, but refrain from getting them to practise at this stage.
- Explain that students will hear two candidates doing the task and they should make notes on what the candidates say. Play recording 5.1 while students take notes.
- Ask students to compare their notes in pairs before giving feedback to the whole group.

5.1 LISTENING SCRIPT

1

Candidate 1: The reason why I don't like this type of weather is that it saps all your energy, you know, you just feel like lying around the house all day, you feel really listless and you just don't get anything done. Either that, or you freeze to death! You see, if I'm at my parents' house, they always have the air conditioning turned up really high and it's just so cold. The other thing is we have this type of hot, humid weather all year round – the only difference is between the dry and the rainy season, but it's always hot and very sticky – really not very nice at all.

2

Candidate 2: One of the things I love about this hot sunny weather is feeling the sun on my face, you see in my country the winters are long and very, very cold, so it's always a relief to have some sunshine. I like going walking in the country with my husband. We both love nature and really enjoy the countryside when it's green and beautiful. The sunshine gives me energy and makes me feel happy. It's rarely too hot in my part of the country, so we just enjoy it as we know it's not going to last too long!

B ▸ First ask the class if they can remember what phrases each speaker used to introduce their explanation. Don't worry if they can't remember; they will hear them again.

▸ Play recording 5.2. This time students listen for the specific language used to introduce the explanation part of the card. After the recording, elicit the phrases from the class.

ANSWER KEY

1 The reason why I don't like … is …

2 One of the things I love about … is …

5.2 LISTENING SCRIPT

1

Candidate 1: The reason why I don't like this type of weather is that it saps all your energy, you know …

2

Candidate 2: One of the things I love about this hot sunny weather is feeling the sun on my face, you see …

C ▸ Draw attention to the language in the box. You could drill this language if you wish. Explain how introducing the explanation using this type of language will impress the examiner. Put students into pairs to discuss the kind of weather they prefer and why. Encourage students to use the language provided in the box to introduce their explanation.

3 Speculating

Aims: To introduce the language of speculation useful for answering some of the Speaking Part 3 questions.

To introduce questions in Part 3 which require a speculative answer.

To briefly revise the language of comparing and contrasting (covered in Unit 3).

A ▸ Remind students that they looked at comparing and contrasting information in Unit 3. Explain that they may also need to speculate in their answer and that they should listen out for questions which require them to do this.

▸ Focus students' attention on the questions. Ask them which question requires them to compare and contrast and which one asks them to speculate.

ANSWER KEY

1 speculate; 2 compare and contrast

B ▸ Tell the class that they are going to hear two candidates answering Question 1 in 3A using speculation. Play recording 5.3. Ask students which answer is better, and why.

ANSWER KEY

The first student's answer is better as it uses more language of speculation. (See examples in the language box in 3D.)

5.3 LISTENING SCRIPT

1

Candidate 1: Er, probably not! Some governments have already made some changes, but in my view it's very unlikely that they will ever do enough. The problem is that, while governments might introduce green policies to win them more votes, for instance, recycling waste, they probably won't commit themselves to policies that are unpopular with the voters. For instance, getting people out of their cars and onto buses and trains is always a very unpopular measure.

2

Candidate 2: I think it's very important that governments do something about climate change. The earth is getting warmer. The sea levels are rising and the weather is becoming much more violent and unpredictable. We need to do something about this problem before it's too late. There has been a lot of talk about this at international conferences, but nothing very much has been done.

C ▸ Tell the class they will hear the better speaker from the previous recording again. On the second listening, get students to listen for the actual language used to speculate. Elicit this language from the class.

ANSWER KEY

The speaker uses the following language: *probably not; it's very unlikely that; might; probably won't.*

5.4 LISTENING SCRIPT

1

Candidate 1: Er, probably not! Some governments have already made some changes, but in my view it's very unlikely that they will ever do enough. The problem is that, while governments might introduce green policies to win them more votes, for instance, recycling waste, they probably won't commit themselves to policies that are unpopular with the voters. For instance, getting people out of their cars and onto buses and trains is always a very unpopular measure.

D
- Focus students' attention on the language in the box. You could use the language to make some example sentences with the class on the board before moving onto the discussion question.
- Put the students into pairs to discuss the three questions (a–c). Encourage them to use lots of examples of the language of speculation in their answers.
- Round up by eliciting some answers from the whole class. Reinforce the language of speculation by using the board to illustrate answers containing the target language.

Challenge

- If your class isn't having difficulties with the language of speculation, you could introduce other phrases for talking about the future, such as (listed from most likely to least likely to happen):

It's bound to happen.
It's inevitable.
I suppose it's a distinct possibility.
There's a (remote/outside) chance that it might happen.
I don't think there's a remote possibility.
It'll never happen in a million years.

express tip

- Read out the information in the box while students follow in their books.

4 Communicating your ideas clearly

Aims: To show how students can make themselves more easily understood by using signpost words to introduce ideas and examples.

To provide an opportunity for students to hear candidates answering Part 3 questions, and to practise answering themselves.

A
- Explain that students can express themselves more clearly by giving some indication of what they are going to say before they say it. This will help the examiner follow their ideas more easily.
- Focus students' attention on the example *for instance*. Ask students to speculate how the sentence might continue, then explain that the words *for instance* told them what kind of information was coming next.

B
- Explain that students are going to hear four candidates using these signpost words. Point out the example in the table, i.e. *outlining your response*.
- Play recording 5.4 and get the students to complete the table with the appropriate words from the candidates' responses.
- After students have compared their answers, elicit feedback from the class.

ANSWER KEY

1 there are two main problems; firstly; in addition to this
2 for instance; such as
3 as a result; which will mean
4 this is due to

5.5 LISTENING SCRIPT

1

Candidate 1: There are two main problems for the farmers: firstly, I think heavy rain causes the soil to be washed away, I mean the soil with all the minerals in it. In addition to this, many plants rot in the ground because they never have time to dry. Think when it rains very heavily, for example, in my country, it is very difficult for the farmers.

2

Candidate 2: There are things we can do to help slow global warming. For instance, we can stop using up all the coal and gas, and use cleaner energy such as power from the wind or sun.

3

Candidate 3: Rising temperatures will lengthen the food-growing season in some countries. As a result, farmers will be able to produce more crops, which will mean less hunger in the world. I'm sure you'd agree this is a good thing … er, a benefit.

4

Candidate 4: It's generally agreed that the earth is getting warmer, this is due to more and more greenhouse gases in the atmosphere.

C

- Ask students to look at the Part 3 question in the box and the four answer beginnings which follow.
- Tell students to complete the sentences with a signpost word and their own ideas before comparing with their partner. You could get students to practise the questions and answers in pairs and see how well they can use the signpost words in their responses.

5 Individual long turn

Aim: To provide an opportunity for students to practise Part 2 of the Speaking exam, and to consolidate the skills introduced in this unit.

for this task

- Ask students to read the information in this box individually. Answer any queries they might have.

EXAM PRACTICE

- Put the students into pairs and assign roles of candidate and examiner.
- Student A should prepare notes for one minute before speaking about the topic for 1–2 minutes uninterrupted. When students are making notes, encourage them to write down keywords, rather than full sentences. Student B should listen carefully, especially for examples of introducing the explanation part of their answer using appropriate language. After 1 or 2 minutes, Student B can interrupt and ask a few follow-up questions. Make sure that the follow-up questions are answered briefly.
- Ask Student B to give feedback to the candidate on his or her performance, focusing particularly on the points in the *for this task* box.
- Students then change roles and repeat the activity.

6 Two-way discussion

Aim: To provide an opportunity for students to practise Part 3 of the Speaking exam, and to consolidate the skills introduced in this unit.

for this task

- Ask students to read the information in the box individually. Answer any queries they might have.

EXAM PRACTICE

- Put the students into pairs and again assign roles of Student A and Student B (candidate and examiner).
- Student A should close his or her book. Student B should interview using the questions on page 59. Student B must listen carefully to Student A, paying particular attention to the language used to express and justify opinions, and language of speculation.
- Ask Student B to give feedback to the candidate on Student A's performance, particularly focusing on the points in the *for this task* and *express tip* boxes.
- When students have finished, ask them to change roles and repeat the activity.

express tip

- Read the information in the box while students follow in their books. Reinforce that it is very important for students to give extended answers to all questions – even those that could be answered with just *Yes* or *No*.

IELTS Express Speaking DVD

- If you are using the *Speaking DVD* which accompanies *IELTS Express*, Section 3 – Part 2 of the DVD relates to the content of this unit. The DVD could either be shown at the beginning of this lesson to give students a general idea of the content and format of Part 3 of the IELTS Speaking exam, or alternatively, it could be shown at the end of the lesson to provide a recap and demonstration of the material covered in this unit.

For more information on the *IELTS Express Speaking DVD* and how to integrate it into your lessons, see page 109.

Globalisation

LISTENING

Section aims:

- To highlight the importance of listening and writing simultaneously, and to give practice in this skill.
- To raise awareness of the use of distractors in IELTS listening texts.
- To show how a speaker's meaning may be implied rather than explicitly stated.
- To introduce and give practice in Listening Section 3: Academic Dialogue.
- To practise a number of task types: notes completion; sentence completion; classification.

1 Introduction

Aims: To introduce the unit topic of globalisation and to elicit some of the vocabulary which arises later in the unit.

To give students the opportunity to briefly discuss some key aspects of globalisation.

A

- Before students open their books, ask them to think about what clothes they are wearing. Ask them: *Do your clothes have a brand name? Do you know where they were made?* Maybe get them to check the labels. Direct students to tell their partners about their clothes and where they were made. Ask one or two students why they chose to buy a particular brand and not another.
- Ask students to open their books and look at the picture at the top of page 60. Ask a student to describe what they see. Ask the class: *What does this photo show? What are these people doing?* (the photo shows the Chicago Board of Trade; buying and selling stocks, shares, etc)
- Ask students to discuss the questions briefly in pairs or small groups, then get feedback from the class. Depending on the class, you may wish to discuss as a group from the start. Allow discussion for four to five minutes at most. Go around the class and help with vocabulary. Finally, get feedback from the class.

ANSWER KEY

Suggested answers:

- Globalisation is the process by which national boundaries become less important, leading towards what has been called 'the global village'.
  There are three main aspects of globalisation: *economic* – multinational organisations increase international trade; *political* – exchange of political ideas, creation of economic power blocks, e.g. EU; *cultural* – more foreign cultural artefacts, e.g. films, video games available. These aspects are all closely linked to developments in transport and communications, e.g. the Internet.
- Gap, Starbucks, Microsoft, etc. Forbes.com lists the 500 'best and biggest companies' of 2003 in terms of global sales as: (1) Walmart – American wholesale and retail – $244,524m, (2) TotalFinaElf – French energy company – $107, 673m, (3) NTT – Japanese telecommunications – $87, 962m. Microsoft appear in the list at number 45, with global sales of $28,365m, Gap at number 100 ($13,848m) and Starbucks at number 321 ($3,289m). If you want more up-to-date figures, consult www.forbes.com or similar websites.
- Globalisation has enabled companies to grow as a result of an increase in flow of goods and services, increase in flow of capital and an increase in flow of technology. Globalisation enables similar companies in the same or related industries to pool their resources and experience through mergers and acquisitions, to create large multi-national organisations. Taking advantage of international trade laws, these companies can outsource production to countries with lower production costs and, through careful marketing, can maximise their profits by selling these goods in countries where average income is much higher.

B

- Divide the class into groups of 3–4, and have them discuss the points in the text. Encourage students to explore both sides of the issue. Tell them they do not have to reach a consensus within their group. With group discussions like this, it is a good idea to appoint a discussion leader within each group, whose job is to manage the time and ensure that everyone has a chance to speak and that no one or two people dominate the discussion. Again, give them a time limit of 8–10 minutes. As students talk, monitor and help with vocabulary as required.

Extension: *Researching the topic*

Aims: *To give practice in researching, summarising and presenting findings.*
To provide a listening opportunity for other students in the class.

Preparation

- You may like to do some preliminary research yourself on this subject, maybe printing off some of the articles from the websites mentioned below, especially if your students do not have access to the Internet, or you wish to check if the texts are appropriate for your particular class.

Procedure

Researching a topic, summarising and presenting findings is an important part of university life, so this exercise is particularly relevant for university-bound IELTS candidates.

- Divide the class into five groups and assign one of the five consequences in 1B to each group. (With a class of fewer than five students, select the most suitable consequences for your class.)
- Ask students to research each consequence for homework and prepare a short talk of maximum three minutes on the advantages and disadvantages of each. For further reference see the following:
  www.globalisationguide.org gives an overview of arguments both for and against such aspects of globalisation.
  Another good starting point is:
  www.guardian.co.uk/globalisation/archive – in particular the following articles: *The use of low pay economies; Why poor people prefer to be protected from free trade; Is it all bad when UK jobs go to India?; Globalisation: good or bad?*
- Before they give the talk, each group should be given time to pool their ideas. You can decide how formal the presentation should be. But be prepared to take notes and give feedback on all the things that the presenter did well (e.g. eye contact, volume of speech, use of signpost words) and those areas that they might consider in future (e.g. lack of summary, pronunciation of certain words, spelling on acetate or in PowerPoint, etc).

IN THE EXAM

- Draw students' attention to the *IN THE EXAM* box. Ask students to read the information in the box. More information on classification questions is provided later in the unit.

2 Listening and writing simultaneously

Aims: To encourage and give practice in the skill of listening and writing simultaneously.
To introduce and give practice in notes completion questions.

A

- Ask students to look at the set of notes and read through Questions A, B and C on page 61 – this will show them how this exercise unfolds.
- Tell students they will now hear the first part of a recording. Ask them to identify the answer to Question 1 but tell them not to write it down yet.
- Play recording 6.1. Check with students that they all have the answer.
- Now tell students they will hear the second part of the recording. Tell them to write the answer to Question 1 while they listen for the answer to Question 2.
- Play recording 6.2, then check Question 1.
- Tell students they will now hear the rest of the recording. Tell them to write the answer to Question 2 while they identify the answer to Question 3 and to continue to listen and record their answers for the remaining questions. Play recording 6.3.

ANSWER KEY

1 Future Is Bright; **2** Dr Jack Jones; **3** Black Books; **4** 1999; **5** 2nd

6.1 LISTENING SCRIPT

(T = Tutor; B = Brad; C = Janet)

T: OK. Before we all rush off on our holidays, I need to give you your reading assignment for the break.

B: Oh, no ... reading.

T: I'm afraid so, Bradley. Now, as you may be aware, there have been literally thousands of books published on the subject of globalisation. Some of the authors have become quite famous; Naomi Klein, for example, with her book *No Logo*, which became an international bestseller.

J: Oh, I've heard of that, I think my brother's got a copy. Maybe I could borrow it.

T: Yes. Well, why not? It's a fairly easy read. Another author, the Peruvian economist, Hernando de Soto, has some very interesting ideas on ways of eradicating poverty in third world countries by freeing up the assets tied up in buildings on land without any clear ownership. He's very influential and has worked closely with the likes of Bill Clinton in the past.

B: So, do you want us to read his book?

T: Not just yet, no. The first book I'd like you to read is called: *The Future Is Bright, the Future Is Global.*

6.2 LISTENING SCRIPT

Tutor: They used to have a copy in the library, but I think it's disappeared. It's written by Dr Jack Jones, who used to lecture at this university.

6.3 LISTENING SCRIPT

Tutor: It was first published by Black Books, who are specialists in this kind of thing, in 1999. But please make sure you buy the second edition. The world changes so quickly these days. Now the other book is against globalisation. It's called *Hands off the Planet* by Ted Crilly. It's published by Craggy Press and ...

3 Identifying distractors

Aim: To raise awareness of the use of distractors in IELTS listening texts, and to help students avoid selecting distractors as the answer.

A

- Direct students' attention to the sentence completion task in the box. Tell students that with some task types, it helps to identify exactly what question is being asked. Ask students: *When completing the gap in this sentence, what question are you answering?* (Who or what from wealthy countries does Alison think WTO rules do not favour?) Write this question up on the board.
- Ask students to identify keywords in this question (*wealthy countries, Alison, WTO rules* and *not favour*). Ask if they can think of any synonyms for these (rich/well-off nations; World Trade Organisation/regulations/code; not support/not encourage). Ask students to identify the type of answer needed here (a noun such as an organisation or a group of people, an object or an activity, e.g. *imports or importing*).
- Finally, ask for predictions for the answer but do not confirm any just yet.

B

- Ask students to cover up the extract from the listening script. Explain that they are going to hear the first part of a conversation between two students.
- Play recording 6.4 and ask students to answer Question 1. Check the answer as a class, but don't explain why the answer is correct at this point.

ANSWER KEY

smaller companies

6.4 LISTENING SCRIPT

(A = Alison; D = Dave)

D: Hi there, Alison?

A: Hello, Dave. Long time no see. What've you been up to?

D: Oh, this and that. Research mainly.

A: Researching what?

D: The WTO.

A: World Trade Organisation?

D: Yes. It's all part of this project I'm doing on globalisation.

A: Oh, yeah. We did that last year. What do you make of it then?

D: Well, it's not exactly the most caring of organisations, is it?

A: What do you mean, Dave?

D: WTO rules favour the larger companies from wealthy countries.

A: In what way?

D: Well, by prohibiting protection through discriminatory tariffs, it's hard for poor countries to build up domestic industries.

A: That may be the case. But I'm sure that's not a deliberate policy. Anyway you could argue that the rules laid down by the World Trade Organisation don't exactly help smaller companies from the richer nations either.

D: Why not?

A: Well, many companies in wealthy countries, especially textile and clothing producers, oppose globalisation because they can't compete with cheaper imports made in countries with lower production costs.

D: ... like China.

A: Exactly.

C

- Students now have the chance to analyse an extract from the listening script, identify distractors and check their answers. Direct students to follow the instructions in their book and answer the questions.
- Ask: *How many students got the right answer, 'smaller companies'? How many put 'larger companies'? Why this is the wrong answer?* (this is what Dave says, not Alison) *What is the distracting information?* (Dave uses almost exactly the same words as the sentence completion task, but he is the wrong speaker.)

ANSWER KEY

Alison: What do you mean, Dave?

Dave: ~~WTO rules favour the larger companies from wealthy countries~~.

Alison: In what way?

Dave: Well, by prohibiting protection through discriminatory tariffs, it's hard for poor countries to build up domestic industries.

Alison: That may be the case, but I'm sure that's not a deliberate policy. Anyway you could argue that the rules laid down by the WTO don't exactly help (smaller companies) from the richer nations either.

D

- Now repeat the process with Questions 2 and 3. Direct students' attention to the two sentences. Ask them to identify the question, keywords, synonyms and type of answer required. Have students try to predict the answer.
- Play recording 6.5. Students listen and record their answers.

ANSWER KEY

2 weakens; **3** seed companies

6.5 LISTENING SCRIPT

(A = Alison; D = Dave)

D: Ah. And that's another thing!

A: What is?

D: Democracy. The WTO isn't the most democratic of organisations, is it?

A: Why do you say that? You know, all of the WTO's rules have to be ratified by member states and all decisions are reached through consensus.

D: Yes, but all those decisions are made behind closed doors.

A: Maybe, Dave. But I still believe that the WTO is a force for strengthening democracy throughout the world, as it encourages international trade and therefore the exchange of ideas and beliefs, including democracy.

D: I can't see how you arrive at that conclusion, Alison. No, if you ask me, it's quite the opposite. The WTO actually weakens the democratic process because it allows the formation of enormous multinational organisations that are richer and more powerful than some countries. And that can't be good. When it comes to global democracy, the WTO has a weakening effect.

A: I suppose you're going to tell me that the WTO should regulate international companies over pollution next.

D: And so they should. The WTO allows global companies to locate pollution-producing industries in poor countries.

A: This idea is nonsense, Dave. Why would a company choose to relocate a whole plant to the other side of the world. The cost would be enormous. It would be much cheaper for the company to clean up the existing plant.

D: Maybe, but look at the extensive logging of the rainforests, Alison. You must agree that the WTO should regulate that?

A: The WTO's regulations allow for countries to protect such natural resources. What does worry me is the way agricultural seed companies focus on high-yield, disease-resistant plants at the expense of other plants. This policy is destroying plant biodiversity and that can only spell trouble. No, these seed companies need regulating.

D: Well, at least we're agreed on something. Fancy a cup of coffee?

A: Only if it's Fairtrade.

D: What? In this place? You'll be lucky. Come on ...

E

- Ask students to turn to listening script 6.5 on page 124. Ask them to identify the distractors, the answer and any information which led them to the correct answer.

Background on WTO: The World Trade Organisation (WTO) is an international organisation, which oversees a large number of agreements defining the 'rules of trade' between its member states. The WTO is the successor to the GATT (General Agreement on Tariffs

and Trade) that was set up in 1947, and operates with the broad goal of reducing or abolishing international trade barriers. It ensures trade among nations operates smoothly, freely and orderly. In the late 1990s, the WTO became a major target of protests by the anti-globalisation movement.
(Source: http://en.wikipedia.org/wiki/globalisation)

4 Understanding meaning

Aims: To raise awareness of the use of indirect language and how meaning can be implied indirectly.

To raise awareness of the importance of intonation in conveying meaning.

To introduce and give practice in classification questions.

A

- Tell students they will hear six short listening extracts. Ask them to listen and decide whether statements 1–6 are true or false.
- Give them a few seconds to read statements 1–6, then play recording 6.6. You could do Question 1 as an example, checking answers before playing the rest of the recording.
- Ask students to check answers against the listening script on page 125. For Questions 5 and 6, students may not know the answer. Don't reveal the answers, just play extracts 5 and 6 one more time. Check answers and ask students how they know.

ANSWER KEY

1 T
Note 'not unacceptable' (a double negative, meaning the opposite)

2 F
Note 'it's not unheard of' (a double negative, meaning the opposite)

3 F
Note The speaker employs two idiomatic expressions, 'on the table' and 'at the end of the day'.

4 F
Note 'I couldn't agree with you more' means 'I agree with you a lot' not 'I disagree'.

5 F
Note upward inflection on 'Oh, great' (happy)

6 F
Note downward inflection on 'Oh, great' (unhappy, denoted by the use of sarcasm)

6.6 LISTENING SCRIPT

1
A: Are the British happy with the conditions as laid out in the contract?
B: Well, they said the terms were 'not unacceptable'...

2
A: And do you actually believe that the French would agree to such a deal?
B: Well, it's not unheard of.

3 While the offer on the table is far from perfect, at the end of the day it's almost certainly the best we are going to get.

4
A: Mr Yamamoto, these designs are just what we need.
B: I couldn't agree with you more, Señor Ramirez!

5
A: Mr Edwards, the Chinese have withdrawn their offer.
B: Oh, great!

6
A: Mr Edwards, the Japanese have withdrawn their offer.
B: Oh, great!

Extension: *Understanding British understatement and coded speech*

Aim: *To introduce students to the (mainly British) habit of using evasive language, especially in the context of business meetings.*

Preparation

- Prepare a worksheet using the examples over the page. Either give students the two lists of words presented as a matching exercise, or give them the explanations and read out the 'British' expressions.
(Note that the sentences are matched to show you the answers; you will need to jumble them up.)

Procedure

- The British often use understatement in speech in order to minimise any emotional charge. This tool is particularly popular with politicians and those in business management. For non-British speakers of English, this habit can be confusing. Here is a brief introduction of some of the more common expressions and their 'translations' (adapted from www.crossculture.com).
- In pairs or groups, ask students to match the two columns.

As a follow-up, ask students to work in pairs or small groups and script short dialogues using these expressions.

Following in two columns:

When the British say:	They really mean:
1 That's a good question.	**a** I don't know the answer.
2 It's not bad.	**b** It's actually pretty good!
3 We'll certainly consider that.	**c** That is totally unacceptable.
4 Let me make a suggestion.	**d** This is what I've decided to do.
5 I agree, up to a point.	**e** I disagree.
6 We'll have to review your position.	**f** You're going to be fired.
7 That might be just a bit tricky.	**g** It's impossible.
8 Hm ... that's an interesting idea.	**h** I can't reject it outright.
9 It's not for me to influence you either way but ...	**i** This is what I strongly advise you to do.
10 I'll call you.	**j** I won't call you.

B

- Tell students that a common IELTS task type is classification. In these tasks, candidates are often required to match an opinion with a speaker. Remind students that as they have just seen, speakers often express their opinions in an indirect manner and so it is important that they listen carefully for such things as indirect structures, double negatives and inflection. In other words, they need to be able to 'read between the lines' and identify what is *meant* rather than just what is *said*.
- Direct students to the classification task in the box. Ask them to skim through it.
- Focus students' attention on Question 1. Spell out exactly what it is they have to do: *You have to identify who has this opinion. Is it just Peter? Just Katya? Or both Peter and Katya?* Ask students to read the bulleted questions below the box.
- Tell students they will now listen to two students discussing globalisation. Play recording 6.7. Students listen, answer the question and then discuss with a partner.
- Elicit feedback from the class, checking answers. Replay the section as necessary.

ANSWER KEY

- Yes

Note 'I'd say the Internet was really important in terms of globalisation.'

- Katya agrees

Note 'I think you've got something there ...'.

- The answer is C.

6.7 LISTENING SCRIPT

(K = Katya; P = Peter)

K: OK, Peter, we need to decide about our presentation next week.

P: OK, Katya. What do you think we should talk about?

K: Well, Dr Chobham said to look at some of the factors that have contributed to the process of globalisation.

P: Er, yes. I was thinking maybe we could do something on the Internet.

K: Really?

P: Yeah, along with things like satellite TV and cheap flights, I'd say the Internet was really important in terms of globalisation.

K: I think you've got something there, Peter. I mean anyone can get hold of all that information, anywhere on the planet.

P: All you need is a computer, a modem and a phone line.

K: Precisely.

express tip

- Read the information in the box while students follow in their books. Remind students that distractors may use the exact wording in the question, rather than a synonym or paraphrase.

C

- Tell students they will now hear the rest of the conversation. Ask them to listen carefully and answer Questions 2–5.
- Play recording 6.8.
- After listening, ask students to check their answers with a partner, then get feedback from the class.
- Play the recording again, stopping after each answer.

ANSWER KEY

1 C; **2** A; **3** B; **4** A; **5** A

6.8 **LISTENING SCRIPT**

(K = Katya; P = Peter)

K: And apparently, I was reading, the Internet and mobile telephones allow developing countries to leapfrog steps in the development of their infrastructure.

P: What does that mean?

K: Well, for example, the Philippines has a poor landline telephone system, but with a mobile phone and computer, you don't need to use it.

P: ... I don't even know anyone here who uses a computer with their mobile!

K: But in my book, the Internet has moved too far from its non-commercial roots. When it was created, it was meant to be a tool for people to communicate with each other. These days it's dominated by big business which is only interested in selling you yet more stuff. I get so much junk mail, and all those pop-ups!

P: Oh, that doesn't bother me. I rather like to know what's on offer.

K: The Internet could also be seen as divisive.

P: In what way?

K: Well-off countries have much greater access to the Internet and communication services in general. What we are witnessing is an information revolution and less well-off countries are getting left behind.

P: Up to a point. Yes, not everyone has access to the Internet at home. But many places have shared communal access – some villages in Africa, for example. But on the whole, it's such a great way of exchanging ideas.

K: Ha! I think you'll find it's a one-way street. The vast majority of websites are in English and western values dominate.

P: I know, I know. You think it's a kind of cultural imperialism.

K: I think that's a fair assessment, don't you?

P: I think you're exaggerating the situation there, Katya. For me, and millions of other people, it's just an easy way of keeping in touch with family and friends, even when you are thousands of miles away.

K: Ah, that reminds me. It's my mum's birthday today. I forgot to send her a card!

P: Why not send her an electronic card?

K: Great idea! Where would we be without the Internet?

5 Classification

Aims: To give students the opportunity to consolidate the skills learned in this unit.

To present further practice in answering classification questions.

for this task

- You could ask students to close their books and describe classification tasks to each other. Ask them to describe what a candidate should do before and while they listen. Ask them how they should record the answers. Now get students to open their books and underline the key information in the *for this task* box.
- Tell students they will now do another classification exercise. Remind them to follow the procedure in the *for this task* box.

EXAM PRACTICE

Questions 1–4

- Play recording 6.9.
- When the recording has finished, check answers as a class. Alternatively, carry on with Section 6, which is a continuation of this conversation, and check answers later.

ANSWER KEY

1 A; **2** B; **3** B; **4** B

6.9 **LISTENING SCRIPT**

(T = Tutor; B = Brad; J = Janet)

T: Welcome back. I trust you had a good break and that you managed to read the books I recommended to you. Any problems, Brad?

B: You know, I thought *Hands off the Planet* might be difficult to get hold of. As it turned out, they had a whole stack of them in my local bookstore. It was even on special offer!

J: Yeah, and you get a free password to enter a website dedicated to the book.

T: Really, Janet?

J: Yeah, I tried to take a look at it but the link wasn't working.

T: Ah well, and what about Dr Jones' book?

B: The bookstore said it was reprinting at the moment. But in the end I managed to track down a copy in a second-hand bookshop.

T: Smart thinking there, Brad. How about you, Janet?
J: Well, my brother had a copy so I just borrowed that.
T: Good, so what did you make of them?
B: I loved *Hands off the Planet*, it was such an easy read, unlike *The Future Is Bright*. I mean, it kept losing me, the argument just kept jumping around.
J: I know what you mean. It wasn't helped by the fact that quite a few of the quotes in foreign languages were left untranslated; it's as if we're all expected to be multilingual!
T: Yes. I'm afraid that Dr Jones does like to show off his familiarity with different languages.
I'll certainly make that point to him next time I see him.
J: But I think the main problem with Dr Jones' book was that it assumed a previous knowledge of the subject.
B: Yeah, right. There were some chapters where I felt way out of my depth. I'd no idea what he was talking about.
J: I had to get my brother to explain it to me!
B: I just didn't feel Dr Jones' book was very user-friendly. Unlike *Hands off the Planet*, it had no illustrations and the section containing the extended interviews with all those foreign businessmen just went on and on.
J: Didn't it just!
T: Well, it is a little on the long side, yes, but I think it remains a relevant and valuable resource, though on reflection it may have been a wiser option to have put these in the back of the book.
J: As an appendix?
T: Precisely.

6 Sentence completion and notes completion

Aims: To give students the opportunity to consolidate the skills learned in this unit.
To present further practice in answering sentence and notes completion questions.

for this task

- Ask students to read the *for this task* box, and answer any questions they might have.

EXAM PRACTICE

Questions 5–8

- Tell students they will now hear another conversation about the books. Ask them to read the rubric and Questions 5–8.
- Remind students to follow the steps in the *for this task* box before and after they listen.
- Play recording 6.10.
- Ask students to check answers in pairs and then as a class. Alternatively, play the recording again in sections to confirm answers.

6.10 LISTENING SCRIPT

(T = Tutor; B = Brad; C = Janet)
T: So you preferred *Hands off the Planet*, did you, Brad?
B: Yeah, I thought it was really interesting. Crilly obviously spent an awful lot of time preparing this book, all those amazing facts and figures. The chapter on how cinema, TV and newspapers are becoming more global was really well researched. In fact, I was shocked to read just how powerful and influential some of these media corporations are.
J: Yes, though I thought at times the author just conveniently overlooked any data that didn't support his argument. It seemed to be quite biased, I thought.
B: That's because he's passionate about all this. He's very concerned about the future of the planet.
J: Well, that's highly commendable, I'm sure.
But oversimplifying things to such an extent greatly distorts the true picture, and by adopting so radical a position, he can actually put people off.
B: Yes, but he sees it as his mission to make people sit up and take notice.
J: Well, to be honest, I'm surprised we were asked to read this book at all.
T: Really, Janet? What makes you say that?
J: It's quite lightweight, isn't it? I'm not surprised they had so many in the bookshop. I don't know, I just didn't find the tone academic enough for a serious study.
B: You mean you didn't like the Captain Planet comic strips? I thought they were hilarious!
J: Yes, I liked them. They were quite amusing.
But I didn't think that they were particularly appropriate for a serious subject such as globalisation.

T: Hmm, I tend to agree with you there, Janet, but other students have read it in the past and most of them have been favourable towards it.

J: Another factor which I felt detracted from the academic nature of the book was that there was no index, whereas the one in Dr Jones' book is excellent.

B: Ah yes, it was superb. More than could be said for the bibliography in *Hands off the Planet*. It's virtually non-existent.

T: Well, maybe there's some more information on the website, if you can make it work that is. OK, thank you for your comments. All very interesting and most useful. Now if we could just focus in, on some of the ideas expressed in these books ...

express tip

- Read the information in the box while students follow in their books. Remind students of the various techniques for expressing opinion they encountered in Section 4.

EXAM PRACTICE

Questions 9–12

- Repeat the steps above for Questions 9–12.
- Play recording 6.11.

Challenge

- Play recordings 6.10 and 6.11 back to back and ask students to answer Questions 5–12. Then check the answers as a class.

ANSWER KEY

5 (really) well researched; **6** support his argument; **7** academic; **8** bibliography; **9** Right or Wrong; **10** 1,500; **11** February 2nd; **12** internal post

6.11 LISTENING SCRIPT

(T = Tutor; B = Brad; C = Janet)

B: ... especially with bananas and so on.

J: Oh, I know exactly what you mean.

T: OK. Time for us to wrap up. Now, I'd like you to write an essay. The title is ... 'Globalisation: right or wrong?'.

B: How many words?

T: One thousand, five hundred.

J: When is it for?

T: Where are we? January 21st. Shall we say in one week's time? The 28th?

B: Oh. I'm not sure, we've got exams till the 26th.

T: Fine. Let's call it February 2nd. That will give you the weekend.

J: OK. And do you want us to email it to you?

T: Best not to. We've had a few problems with the system in the past. No, pop it in the internal post. Right, I'd better hurry. I've got a lecture in five minutes, now where did I put my gloves?

express tip

- Read out the advice in the box while students follow in their books. The ability to write and listen simultaneously is essential for IELTS success and a skill that students will need to practise.

6

WRITING

Section aims:

- To show students how to describe trends as well as specific data in graphs and tables.
- To introduce the skills for describing a process.
- To introduce and provide practice with both task types in Academic Writing Task 1.

1 Introduction

Aims: To introduce the topic of mobile phones and to elicit some of the vocabulary which arises later in the unit.

- Draw students' attention to the photographs. Ask the class what the people are doing in each photo.
- In pairs or small groups, ask students to answer the bullet point questions. If your students are interested in technology, you could discuss what the next generation of mobile phones might be able to do, or what they would like them to be able to do.

IN THE EXAM

Draw students' attention to the *IN THE EXAM* box. Have the students read the information in the box one paragraph at a time. After Paragraph 1, ask students: *What is not required in Task 1?* (your opinion outside the information given) After Paragraph 2, ask students: *What kind of diagram could you be given?* (table, pie chart, line chart, bar chart or a combination); *What information should you describe?* (key trends and differences – though not all of them). After Paragraph 3, ask students: *What other type of task can IELTS Task 1 be?* (a process or procedure) Tell them that they will practise one of these later in the unit.

2 Describing trends

Aims: To give practice in analysing the key features of graphs.

To practise identifying trends and to introduce useful language for describing them.

A

- Focus students' attention on the graph, telling them to cover up the questions. Set a time limit of about thirty seconds and ask them to focus on the key information.
- Tell students to close their books and tell their partner everything they can remember about the graph, such as the title, axes labels. This is the information they should look at first.
- Tell students to open their books again and go through Questions 1–8 with a partner. Go around the class and prompt students as necessary. Get students to swap partners to check answers. To speed up feedback, write the answers on an overhead projector or on the board.

Support

- If students are finding this task difficult, you could write the answers mixed up on the board or overhead projector.

ANSWER KEY

The line graph shows the change in the share price of Nokia in US dollars between March 2004 and March 2005.

1 price of each share in US dollars

2 one year from March 2004 to March 2005

3 past tenses (mostly past simple) because the period has finished

4 most important: March 2004, August 2004

5 March 2005 is lower than the previous year

6 a decline in share price

7 an increase in share price (This is positive for Nokia since the share price has increased from its trough in August.)

8 March — April saw a peak, then a general fall with some levelling off until August; August onwards — a general increase until December; some fluctuations with downward movement until February; February to March recovered (back to December levels).

B

- Draw students' attention to the Language Bank on the back cover flap. Point out the part of speech in each case (noun, verb, etc). If it is a verb, elicit the base form and past tense, e.g. *fall – fell*. To make this language focus more active, you could get students to come out to the board and draw what each one would look like on a line graph to ensure students are clear on the meaning.
- Do Question 1 together as a class. Point out that the answer must contain a noun to follow *there was*.
- Get students to do Questions 2–4. In each case, remind students to consider meaning (according to the graph), but also grammar (the correct part of speech to fit the space). Check answers as a class.

▸ Point out the time phrases used in 1–4 and the appropriate prepositions: *in, between* x *and* y, *from* x *to* y, and get students to highlight them.

ANSWER KEY

Suggested answers:

1 a peak and then general fall with some levelling off until August

2 fell overall

3 there were some fluctuations but the price remained fairly constant/there was little change

4 there was a gradual increase/upward trend/rise

express tip

▸ Read out the advice in the box while students follow in their books. The previous exercise focused on general trends and the language for describing them, but it is important for students to note that specific data/figures must also be quoted, or students will lose mark on the task achievement criterion. Exercise C will draw attention to this.

C ▸ Focus students' attention on the Motorola graph. Ask students: *What is similar or different from the Nokia graph?* (similar: labels on axis, same time period, same type of information; different: a different phone company and therefore different trends over the same period).

▸ In pairs, get students to answer Questions 1–5 and 8 from Exercise 2A, but this time for the Motorola graph. During feedback, point out that these kinds of questions are an important first stage to any Task 1 graph. By the time students do the exam, these kinds of questions need to be 'inside their heads' and made part of their routine.

ANSWER KEY

1 price of each share in US dollars

2 one year from March 2004 to March 2005

3 past tense

4 There were a number of peaks in the share price during this period. However, the most significant trough was a fall in price to just $14 a share in mid-August.

5 March saw a lower share price than the same month of the previous year.

8 March to June saw a number of fluctuations, then a significant drop till August; prices rose until October when the price remained more stable until the end of the period.

D ▸ Read the first sentence of the text as a class and ask students to suggest answers for Question 1. Get them to focus on meaning (it went down) and also grammar (it needs to be a noun, possibly with an adjective). Elicit the purpose of this first sentence (it is an overview of the graph) and point out that the first sentence in this task type should always introduce the graph in this way.

▸ Ask students to complete the other gaps individually using the language from the Language Bank. In pairs, have them check that their partner's answer is correct for meaning and grammar.

ANSWER KEY

Suggested answers:

1 a decrease/downward trend/decline/fall

2 several fluctuations

3 sudden increase/rise

4 general downward trend

5 fell to the lowest point

6 rise/increase

7 reached a peak

8 downward

Challenge

▸ For early finishers or stronger classes, get students to suggest as many options as possible for each gap. This will give them the opportunity to experiment with the language and encourage them to expand their vocabulary.

E ▸ The preceding exercise is a hidden model answer and the real task is not just in completing the description of the Motorola graph, but also analysing this model so that students are better able to write the Nokia one. Remind them of the overview at the beginning and the one-sentence summary at the end that draws on the data given only. Before students start writing, elicit the important points covered in the preceding exercises: writing an overview and short summary at the end; including specific figures and data to support your points as well as general trends. Remind them of the information they have from Exercise 2A on the Nokia graph and the Language Bank at the back of the book.

▸ If you are short of time, this could be done for homework.

Extension: *Describing trends*

Aim: *To provide additional practice describing trends.*

Preparation

- Source information on market trends. e.g. www.bbc.co.uk (go to *business* or *market data* and type in a company name and a time period). This could be other mobile phone companies (e.g. Samsung) or you could find information on other companies such as Apple and IBM over a more recent period. Print out two graphs, labelled A and B.

Procedure

- Put students in pairs. Give half the pairs Graph A and the other half, Graph B. In pairs, students write questions for the graph they have been assigned, similar to those in 2A on page 64.
- When they have finished, check the questions are correct and appropriate for focusing on the details. Ask students to pass the graph and their questions to a pair who were working on the other graph. Once the questions have been answered correctly, they can then write a full answer to the graph.

3 Describing a process

Aims: To teach the skills of identifying stages in a process.
To introduce the language of time, order, reason.

A

- Ask students to close their books. Explain that Task 1 is often a graph. However, it can also be a process. Elicit examples of a process (making something, e.g. bread; the water cycle; how something works, e.g. a light bulb).
- Write *'Launching a product globally'* on the board and ask what that means. Get students to imagine a company that has a product which is doing very well in their country. *How would the company make this an internationally available product? What would they need to do?* Get students to brainstorm some ideas in groups and do class feedback, e.g. market research; import/export restrictions.
- Open books on page 66 and read the exam question as a class. Ask students: *What does the diagram show?* (the stages a company needs to go through); *Who is this writing intended for?* (a university lecturer)
- Focus students' attention on the diagram and go through any unknown vocabulary.
- Work through Questions 1–4 as a class.
- Point out that these are useful questions to ask yourself when approaching this type of task.

Support

- If you think your class would find the vocabulary here difficult, you could prepare a hand-out with the vocabulary from the diagram in a list on the left. Mix up the definitions in a list on the right. As a lead-in task, get students to match the word to the definition by drawing a line.

ANSWER KEY

1 Since 'Launch product ...' is obviously the end, the process starts with 'Business plan'.
2 6; with sequence linking words (e.g. first, next)
3 present tense – it is a general fact
4 **Suggested answer:** The diagram shows the steps a company should take to successfully launch a new product worldwide.

B

- Tell students to close their books. Write the headings *Time, Order, Reason* and *Example* on the board. Tell students to copy the headings and to leave space below each. Explain that you are going to call out some words and students should put them under the appropriate heading. Give a couple of examples: *first* would go under the heading *Order* and *before* would go under *Time*. Call out the words in the lists in random order, crossing them off as you go to avoid repetition.
- Get students to compare answers and then check by looking at the box on page 66. Check for any student's difficulties with meaning or function.
- Explain that the text in the box is a model answer for the process question but the linking words connected to time, order, reason and example are missing. Do the first question as a class, eliciting why that answer is correct, then get students to do the remaining questions individually.
- Students should check answers with a partner and then as a whole class. Draw attention to the question at the end of this section. The phrases in bold are examples of the passive voice. The passive is usually

used when we do not know who did the action or it isn't important who did the action, or when we want to avoid saying who did it. It is used in processes to focus attention on the subject of the sentence. It has the effect of making something sound more formal and is common in processes. You may feel it is necessary to do a remedial grammar presentation on this.

ANSWER KEY

Suggested answers:

1 In order to; **2** first; **3** Next; **4** Then/After that; **5** such as; **6** Once/When; **7** for example

The passive voice is used when the subject did not do the action. It's usually used to make something sound more formal and is common in processes.

4 Academic Writing Task 1: Report

Aims: To give students the opportunity to consolidate the skills introduced in this unit.

To practise a complete Task 1 Writing task.

To practise writing within a specific time limit.

for this task

- Before reading this, ask students to close their books and write four tips they have learned from this unit for Task 1. Ask students to share their tips with the class. Ask students to read through the points in the box, which prepare them for the following exam practice question. When students have finished reading, get feedback from the class: which ones did they remember and which did they forget?

express tip

- Read out the advice in the box while students follow in their books. Ask students: *Should all your sentences be in the passive voice?* (No, the passive is common in this type of Task 1 question, but it is important to use a variety of structures.)

EXAM PRACTICE

- Read through the question together. Ensure students know what *ATM* and *transaction* mean. This is quite a complex process so it would be a good idea to go through the stages together to check students understand what each represents. Elicit where the process starts (card and PIN entered at local bank).
- Ask students to write the answer in twenty minutes, or if time is short, set it for homework.
- When giving feedback to students on their essays, direct them to the model essay on page 108 of the *Coursebook*. Explain to students that the model essay is not a definitive answer – it is just one way of answering the question.

Model answer

The diagram shows how a transaction works at an automated teller machine in five steps, allowing us to withdraw money from any participating bank in the world.

First of all, Mr Smith inserts his card and PIN into the ATM at a local bank abroad. An electronic message is sent to the central network which is passed on to Mr Smith's own bank in his home country. The local bank will not dispense the cash until it knows the funds are available in Mr Smith's own account.

As soon as the home country bank receives the request and checks the balance, the money is debited from the account. The bank then replies to the message via the network, stating that the local bank can provide Mr Smith with the amount requested. Mr Smith retrieves the money and his card and goes on his way. Later that day a settlement occurs between the two banks facilitated by the central network.

This diagram shows the convenience of ATMs for the user and what goes on behind the scenes to ensure banks only provide money via their ATMs if the account holder has sufficient funds.

Communication

READING

Section aims:

- To introduce the use of distractors in IELTS reading texts, and to teach students a thought process which will help identify them.
- To provide practice in identifying and following arguments, and to show how this is helpful when tackling IELTS questions.
- To practise three task types: multiple-choice questions with single answers; multiple-choice questions with multiple answers; True/False/Not given questions.

1 Introduction

Aims: To introduce the topic of early humans and communication, and to act as a lead-in to the reading.

To elicit some of the vocabulary which arises later in the unit.

To introduce the skill of identifying whether an argument is present in a text.

A
- Direct students' attention to the picture at the top of the page. Elicit answers from the class to questions such as: *What kind of art is this? What does it show? Who do you think painted it? When do you think it was painted? How old do you think it is?*
- Continue the discussion using the questions in the book. Have the students discuss in pairs or groups of three, then get feedback from the class.

B
- Focus students' attention on the two questions, asking them what they think the answers might be. Have them scan the text quickly for the answers, but make it clear that only one of the answers is present in the text. Explain that one of the skills required in the IELTS exam is deciding when a question is or isn't answered in a text and knowing when to stop searching for the answer to avoid wasting too much time.

ANSWER KEY

1 by comparing skull sizes; **2** Not given

Extension: *Internet research*

Aim: *To provide an opportunity for additional topic reading outside the class.*

Procedure

- Allocate the four discussion questions in 1A to students to research in pairs on the Internet (one question per group). Set up feedback as a series of a mini-presentations, which students can give in small groups or to the whole class.

IN THE EXAM

- Draw students' attention to the *IN THE EXAM* box. Point out how it is important to be able to follow the argument being made by the writer. Emphasise the difference between True, False and Not given answers, i.e. a true answer will have the same message as the information in the text whereas a false answer will contain a message which says something which contradicts the information in the text. As in the questions in B in the introduction, an answer that is 'Not given' will probably refer to an idea contained in the text, but will not say the same thing *exactly*. You should also mention multiple-choice questions where you choose several answers from a longer list, as students may not have seen this question format before.

2 Identifying distracting information

Aims: To familiarise students with the purpose of distractors.

To offer a thought process which helps identify distracting information.

A
- This section enhances students' awareness of information which is designed to distract candidates and make them answer *True*, rather than *Not given*. Often questions are designed so that there is information in the text which appears to be the correct answer, but does not actually answer the question. You could exemplify this further by looking back at the questions in 1B, pointing out that both questions are about the origins of language, but only one is answered in the text. Students need to be able

to identify distracting information and resist the temptation to answer immediately what they think to be the correct option.

- Ask students to re-read the text on the previous page more carefully. Explain that once they have identified the correct part of the text through skimming and scanning, they then need to read the section more carefully in order to answer the questions.
- Ask students to answer the question in the box, using one of the options (a–c). They should use the two bulleted questions to help them decide which option is both included in the text and answers the question.

ANSWER KEY

1 Options a and b appear in the text but do not answer the question.

2 Option c appears in the text and answers the question.

B

- In a similar way, there may be options which seem logical but are not mentioned in the text. Again, there is a temptation for students to quickly answer these questions, but they should be careful not to jump at the answer too quickly and read carefully to make sure the answer *is actually contained* in the text.
- Ask students to answer the question in the box, but again, get them to focus on the two bulleted questions, which ask them to look for the option that seems logical and is mentioned in the text.

ANSWER KEY

1 Options **a** and **b** are not mentioned in the text.

2 Option **c** appears in the text and answers the question.

3 Identifying arguments

Aims: To help students follow the main arguments being presented in a text.

To provide practice in identifying if an argument isn't present in the text.

A

- In this section students are going to read a longer text (the rest of the article on the development of language and the human brain) and identify which of a series of arguments are contained in the text.
- Explain that one of the skills required in the IELTS exam is to follow and identify the arguments being made by a writer.
- Focus the students' attention on Questions 1–4, which may or may not be mentioned in the text. Make sure students understand the arguments clearly. Remind students that the 'theme' of the argument may be mentioned, but ask them: *Does it make exactly the same argument mentioned in the text?*
- Ask students to skim the passage on page 76 and decide which points (1–4) are addressed in the text. To encourage skimming rather than careful reading, set a time limit of three minutes for this activity.

ANSWER KEY

1 mentioned

Note 'This view of language is associated with the old idea that logical thought is dependent on words ...'

2 not mentioned

3 mentioned

Note 'The theory sees gesture language as arising originally among apes as sounds accompanying gestures.'

4 not mentioned

B

- This task looks at whether a statement agrees with a point of view presented in the text and therefore introduces students to the idea of Yes/No/Not given questions.
- Ask students to re-read the second paragraph carefully. Then ask students to close their books and, in pairs, paraphrase the main idea of the paragraph to a partner before establishing the idea in class feedback.
- Have them read Statements 1–4 and decide if the statements agree, disagree or state an idea not mentioned in the paragraph. Explain that this idea of identifying whether something agrees, disagrees or is not mentioned, is tested in the Yes/No/Not given questions later in the unit.

ANSWER KEY

1 agrees

Note '... the ability to speak words and use syntax was recently genetically hard-wired into our brains ...'

2 not mentioned

3 disagrees

Note '... the spoken word appeared suddenly among humans ...'

4 agrees

Note '... between 35,000 and 50,000 years ago ...'

4 Multiple-choice questions with multiple answers

Aims: To show how identifying and following an argument is helpful when tackling this task type.
To introduce and provide practice with multiple-choice questions with more than one answer.

for this task

- Ask students to read through the *for this task* box. Explain that the multiple-choice questions are asking them to identify which arguments are contained in the text. They should choose an option if the argument is made in the text *and* if it answers the question. Point out that the options are in the same order as the text and so are very useful in identifying which part of the text the answer must be, i.e. if you have identified Option A as relating to the third paragraph, then you know that none of the other options can relate to Paragraph 1 or 2.

express tip

- Read out the advice in this box while students follow in their books. This tip will help students manage their time more efficiently. If the question is numbered 1–3 then you know there are three marks allocated to the task. Sometimes there may be only one. Point this out to students so that they might decide how long to spend on the question if they are running short of time.

EXAM PRACTICE

Questions 1–3

- Ask students to choose the correct options individually, then check answers in pairs. As they compare, insist that they point to the relevant part of the text which matches with the option. Refer to this information in your class feedback.

ANSWER KEY

The correct options are:

C
Note 'European cave paintings … are seen … as the first stirrings of symbolic and abstract thought, and also of language.'

D
Note 'This view, that spoken language was ultimately a cultural invention …'

E
Note '… there would have been a real need to communicate more effectively and cope with the ever worsening environment …'

5 Multiple-choice questions with single answers

Aim: To introduce and provide practice with two kinds of multiple-choice questions: focusing on a particular section and global (the whole of the text).

for this task

- Go through the *for this task* box with the class, making students aware of the difference in the location of the information for this task and the multiple-option task: in this task the relevant information will be concentrated in one part of the text and will require more intensive reading and understanding on the student's part.

express tip

- Read out the advice in the box while students follow in their books. Global multiple-choice questions are practised in more detail in the *IELTS Express Upper Intermediate Workbook*.

EXAM PRACTICE

Questions 4–5

- First, ask students to identify which part of the text Question 4 refers to. This should be easy to find as the question refers to Michael Corballis. Remind students that the best way to locate names and dates is to scan the text. They should then re-read that part of the text very carefully to choose the correct option.
- Next, ask students to identify which part of the text Question 5 refers to. They should recognise that this is a global multiple-choice question, which doesn't refer to any one specific part of the text.
- Ask students to answer Questions 4 and 5 alone, then check the answers as a class.

ANSWER KEY

4 A
Note 'The theory sees gesture language as arising originally among apes as sounds accompanying gestures.'

5 A

6 True/False/Not given

Aim: To provide practice of True/False/Not given questions.

for this task

- Ask students to read the *for this task* box. Point out that the questions are in the same order as the text, which is very useful in identifying which part of the text the answer is, i.e. if you have identified Question 6 as relating to the third paragraph, then you know that none of the subsequent questions can relate to Paragraph 1 or 2. Go through the bulleted questions carefully, emphasising the difference between 'False' and 'Not given' questions.

EXAM PRACTICE

Questions 6–11

- Ask students to answer the questions individually, then check their answers in pairs. As they compare, insist that they point to the relevant part of the text which matches the option. Refer to this information in your class feedback.

ANSWER KEY

6 True
Note '… analysis of recent evidence suggests we may have started talking as early as 2.5m years ago.'

7 False
Note 'French, for example, clearly does not result from any biological aspect of being French …'

8 True
Note '… the weather took a decided turn for the worse … there would have been a real need to communicate more effectively …'

9 Not given

10 False
Note 'The maximum in brain size achieved by 1.2m years ago …'

11 Not given

SPEAKING

Section aims:

- To help students make their Part 2 talk more engaging through visualising the scene.
- To revise the language function of speculation (from Unit 5).
- To practise evaluating and hypothesising when discussing Part 3 questions.
- To practise Speaking Part 2: Individual long turn and Part 3: Two-way discussion.

1 Introduction

Aims: To generate interest in the topic of languages and to elicit some of the vocabulary which arises later in the unit.

To provide an opportunity for students to practise speculating.

- Ask students to discuss the questions in pairs or threes. The questions naturally require students to speculate while they are discussing. Monitor closely and see how well students are able to do this, as it is one of the speaking functions covered in this unit. When they have finished, get feedback from the class. Students can then check their answers with the answer key at the back of the book.

ANSWER KEY

1 7,000

2 Mandarin: more than 1 billion speakers

3 4,000. See http://www.askoxford.com/worldofwords/wordfrom/vanish for a discussion of dying languages.

4 86%

5 6 million

Oxford's *Ask the experts* site: http://www.askoxford.com/asktheexperts/faq/aboutenglish/numberwords says:

'The Second Edition of the *Oxford English Dictionary* contains full entries for 171,476 words in current use, and 47,156 obsolete words. To this may be added around 9,500 derivative words included as sub-entries. Over half of these words are nouns, about a quarter adjectives, and about a seventh verbs; the rest is made up of interjections, conjunctions, prepositions, suffixes, etc. These figures do not take into account entries with senses for different parts of speech (such as noun and adjective).

This suggests that there are, at the very least, a quarter of a million distinct English words, excluding inflections and words from technical and regional vocabulary not covered by the *OED*, or words not yet added to the published dictionary, of which perhaps 20 per cent are no longer in current use. If distinct senses were counted, the total would probably approach three quarters of a million.'

IN THE EXAM

- Direct students' attention to the *IN THE EXAM* box. This box focuses on Part 3 of the Speaking exam. Emphasise that the examiner is listening for a range of language and that they should try to include examples of speaking hypothetically and speculatively.

2 Visualising the scene

A

- The examiner will listen to a lot of students speaking during his examining session. Explain that making the subject of your talk more vivid will engage the examiner and make your talk stand out from the rest. One way to do this is by visualising what you are describing. Explain that whilst you shouldn't close your eyes in the exam, it can be a useful way of training yourself to really 'see' what you are talking about.
- Have the students close their eyes, relax and get comfortable. Give them lots of time to settle down and wait until there is silence in the class. Act a bit like you are a hypnotist with a calm relaxing voice. Make sure students keep their eyes closed (There will be lots of peeping in the first minute or so!).
- Read the questions on page 81 of the *Coursebook* slowly, in a very relaxing way – keep encouraging students to 'see' the answers to the questions. Give them plenty of time to imagine the scene. When you have finished asking the questions, give students another 30 seconds to continue imagining before gently 'waking them up'.

B

- Put students into pairs and have them work through the questions on page 81, explaining what they 'saw'. Ask them to identify if there were any similarities between their images. Finally, conduct feedback as a class.

3 Individual long turn

Aim: To provide further practice of Part 2 of the Speaking exam.

for this task

- Go through the *for this task* box, emphasising the importance of making notes to help structure the talk and including the important keywords. You should also refer to the *express tip* box, which reminds students to practise as much as they can.

EXAM PRACTICE

- Ask students to make notes to help them respond to the question on the task card asking them to describe the occasion when they found it difficult to understand something in English. Emphasise that they should also spend time explaining whether this was a problem or not and if it was, how they dealt with the situation.
- Put the students into pairs of Student A and Student B (candidate and examiner).
- Encourage Student B to assess the performance of Student A by focusing on the criteria in the *for this task* box as well as the *IN THE EXAM* box on page 13. Student B should also think of one or two brief questions to round off the topic.
- When Student A has finished talking, have Student B give feedback based on the criteria. Then ask students to change roles and repeat the activity.
- Go around and monitor closely during this activity, and try to include points students need to work on in the class discussion at the end.

express tip

- Read out the advice in the box while students follow in their books. The more practice students have before entering the exam, the less likely they are to perform badly because of nerves.

4 Hypothesising, speculating and evaluating

Aims: To revise the language function of speculation covered in Unit 5.

To help students identify which language function is required in answering certain questions.

To provide the language of evaluation and hypothesising, and to practise using this language when answering Part 3 questions.

A

- Explain that in Part 3, the examiner will ask questions to test the candidate's range of language.
- Draw students' attention to the three questions in the box, asking students what they think the connecting theme is between all the questions (one global language).
- In pairs, get students to discuss what language function each question is designed to test. Make sure students are clear on the meaning of *hypothesise*, *speculate* and *evaluate* before asking them to match the questions with the functions underneath.
- Get feedback from the class, focusing on the important language in each structure. Focus on: *Do you think this is a good thing? Do you think it will ...? What do you think the effects would be ...?*

ANSWER KEY

3, 2, 1

B

- Explain that students are going to listen to three candidates answering the questions in 4A. Play recording 7.1 and ask students to complete the gaps.
- Have the class compare answers in pairs, saying which questions from 4A each candidate answered. Highlight the language that is useful for evaluating and hypothesising on the board. Point out how the first sentence gives both sides of the argument before offering a conclusion.

ANSWER KEY

1 on the one hand, then again, on balance
2 if everyone spoke, would be easier
3 would be able

7.1 LISTENING SCRIPT

1

Candidate 1: I think, on the one hand, we can do business more efficiently, but then again, other languages may die out, so on balance I think we need to monitor the situation more closely and make sure other languages survive.

2

Candidate 2: I suppose if everyone spoke the same language, it would be easier to do international business, I mean you wouldn't have to worry about having negotiations in a second language.

3

Candidate 3: In my opinion, political leaders would be able to relate to each other better. You know, sometimes it can be difficult to really connect with someone if you have to speak through an interpreter.

C

- Put students into pairs to discuss Questions 1 and 2. Go around the class and monitor students closely. Listen for examples of students speculating and evaluating. Where necessary, give help to students having difficulties, providing a model of the target language. Choose examples of answers from students who used the language of speculation and evaluation well to model during class feedback at the end of the activity.

Challenge

- If students manage the first two questions without problems, have them practise the language of hypothesising by answering Question 3 in 4A in pairs.

5 Two-way discussion

Aim: To provide practice of Part 3 of the exam.

for this task

- Go through the *for this task* box emphasising that this is where the students can really show off their range of language and gain a higher band score. There is a lot of information in this section, but it is worthwhile going through each point as this box gives an overview of the most important advice. Impress upon students that the examiner is looking for a candidate who appears confident and is able to take the initiative in the conversation.

EXAM PRACTICE

- Put the students into pairs and assign roles of Student A and Student B (examiner and candidate).
- Student B should close his or her book so they cannot see the questions. Student A should interview Student B using the questions at the bottom of page 81. Tell Student B to listen carefully to Student A, paying particular attention to the range of language being used by the candidate. You could also get the examiner to assess the candidate using the criteria set out on page 13 of the *Coursebook*.
- It is not important to use all the questions. In the real exam, the examiner will choose from a number of different topics related to the same theme.
- Ask Student A to give feedback to their partner on his or her performance, particularly focusing on the points in the *for this task* box and the criteria mentioned. Get the examiner to begin his questions with the sentence in the book: *We've been talking about language and communication, I'd like to discuss with you a few questions related to this topic ...* This will familiarise the students with this cue that Part 3 is about to begin.
- As the students are asking and answering, monitor closely for useful language or strong performances and use these in your class feedback.
- When students have finished, get them to change roles and repeat the activity. They can also change partners, and ask and answer the questions again.

IELTS Express Speaking DVD

- If you are using the *Speaking DVD* which accompanies *IELTS Express*, Sections 3 and 4 of the DVD relate to the content of this unit. If you haven't already shown these DVD to your class, it could be shown at the beginning of this lesson, to give students a recap of the content and format of the Parts 2 and 3 of the IELTS Speaking exam.
- At the end of the lesson, you should show your class Section 5 of the DVD, which contains a full model IELTS interview. This will help review everything which has been taught in the course and give students an idea of what to expect in the exam.

For more information on the *IELTS Express Speaking DVD* and how to integrate it into your lessons, see page 109.

Growth and Development

LISTENING

Section aims:

- To raise awareness of features of speech: pace, stress and pausing.
- To apply awareness of features of speech to understanding meaning in listening texts, particularly in relation to summary completion tasks.
- To familiarise students with Part 4 of the Listening exam and to practise a number of task types: short-answer questions, multiple-choice questions with multiple answers, summary completion.

1 Introduction

Aim: To introduce and stimulate interest in the topic of pre-natal development and motor skills development, explored later in the unit.

- Draw students' attention to the three pictures at the top of the page. Ask students how old they think the child is in each picture.
- Put students into pairs or small groups and ask them to discuss the remaining questions in the introduction.
- After a couple of minutes, get feedback from the class, but do not confirm any answers or suggestions just yet. Tell them they will find out the answers during the course of the lesson.

ANSWER KEY

pictures from left to right: 7 months; 19 weeks; 1 to 2 years

Bulleted questions: The answers to this should not be provided to students at this stage. They are in the listening scripts throughout this unit.

IN THE EXAM

- Direct students' attention to the *IN THE EXAM* box. Read out the information in the box while students follow in their books. Remind them that the Listening exam is designed so that the questions generally get more difficult, with Section 4 being the most problematic for the majority of students.

2 Identifying features of speech

Aim: To raise awareness of certain features of speech (namely pace, stress and pausing) and to explore their meaning.

A

- On the board, write the following dialogue.

Sally: How's your sister, Harry?
Tom: The baby's due next week.
Sally: Really? Is it a boy or a girl?
Tom: It's another boy.

- Draw students' attention to the conversation. Elicit the meaning of 'the baby is due' (i.e. the baby is expected to be born). Ask students: *How does Harry feel about the arrival of his nephew?*
- Now ask a student to play the role of Sally. Tell them that you are Tom. Ask the student to act out Sally's part. Ask the rest of the class to listen and decide how Tom feels about his new nephew. The first time, make Tom excited (i.e. read *It's another boy* quickly, as if he is saying *It's fantastic!*). The second time make Tom unenthusiastic, i.e. read slowly, as if he is saying *I don't really care whether it's a boy or a girl.* Depending on your acting ability, you may like to try more feelings. For example, make Tom unsure (i.e. insert a pause – *It's … another boy?* as if he is saying *I think it's a boy. She did tell me, but I wasn't really listening.*).
- Thank the student, then ask the class how they could tell how Tom felt when they heard his voice. Elicit a few responses and then introduce the idea that there are certain features of speech which convey meaning. By way of explanation, read out the introduction to Section 2 while students follow in their books.
- Direct students' attention to the sentences, which identify certain features of speech. Tell them that although the way we speak varies from person to person and region to region, there are certain general truths.
- Ask students to work in pairs to complete each sentence with one word only.
- When they have finished, check answers as a class.

ANSWER KEY

1 pace; 2 stress; 3 pause

B ▸ Note that for this exercise, it's a good idea to have a clean copy of the listening script transferred to an OHP acetate or written on the board to mark the answers so that students can see them clearly.

▸ On the board, write *pre-natal development*. Ask the class what this means (the stages of development a baby goes through before it is born). You could refer the students to the pictures at the top of the page.

▸ Tell students they will now hear a short talk on pre-natal development. They should listen for the three features of speech dealt with in Section 2. Tell them they will start by looking at speed. Direct their attention to the extract of the talk in their books, and make sure they understand what to do.

▸ Play recording 8.1 and ask students to mark up the script to show faster parts, as in the example given. Most students will probably find they need to hear the recording twice.

▸ Ask students to check answers with a partner. Elicit answers and mark up the copy of the text on the OHP or board with your students' suggestions.

▸ Play the recording again, correcting the annotation of the text if necessary.

ANSWER KEY

B and C

For the majority of animals,/and I include human beings in this category,/the most dramatic physical changes actually happen before the organism is born./Is this really the case? Well, think about it!/A mature adult human does not look that different to a newborn baby,/but a newborn baby looks nothing like a fertilised egg!

8.1 LISTENING SCRIPT

For the majority of animals, and I include human beings in this category, the most dramatic physical changes actually happen before the organism is born. Is this really the case? Well, think about it! A mature adult human does not look that different to a newborn baby, but a newborn baby looks nothing like a fertilised egg!

C ▸ Tell students they will now hear the talk two more times. Each time they should mark on a different feature of speech. Ask them to use the annotation as described in the *Coursebook*.

▸ Play recording 8.1 twice more, pausing for thirty seconds or so between each listening so that students can compare answers in pairs. After the final listening, elicit answers from the class, marking up the text on the board or OHP.

▸ Read out the question at the end of this exercise. Ask students to discuss the question with a partner, then get feedback from the class. The basic idea here is that the speaker indicates important information through changes in pace, stress and pausing.

3 Using features of speech

Aims: To consolidate awareness of how features of speech convey a speaker's message.

To explore how such an awareness is useful when answering certain IELTS task types, in particular summary completion.

To introduce and give practice in summary completion tasks and multiple-choice questions with multiple answers.

A ▸ Tell students they will now hear the next section of the talk on pre-natal development.

▸ Refer students to the listening script speech bubble in their books. Ask them to mark up the listening script, using the same annotation as in Exercise 1C, with their predictions for pace, stress and pausing.

▸ Encourage them to read out the speech to each other, noting where changes in pace, pauses and stress naturally occur.

▸ After five minutes or so, play recording 8.2 and ask students to check their predictions. Tell the students that even though the version on the recording may not exactly correspond with their version, that doesn't mean they are necessarily 'wrong'. Different speakers have different rhythms of speech, although sentence stress will be more or less the same.

▸ Annotate the version on the board or overhead projector. Alternatively, have an acetate already prepared.

ANSWER KEY

For that is what we are at the time of our conception/– a tiny, microscopic fertilised egg./Quite amazing, isn't it?/In the following nine months before our birth,/not only do we grow hundreds of times in size,/but we go through three distinct stages of development. The first/as I have just mentioned,/is the fertilised egg./Then around a fortnight after conception,/the egg begins to repeatedly divide itself to become a mass of cells,/known as the embryo./Two months later, this embryo has grown to approximately 2cm long and is referred to as a foetus.

8.2 LISTENING SCRIPT

For that is what we are at the time of our conception – a tiny, microscopic fertilised egg. Quite amazing, isn't it? In the following nine months before our birth, not only do we grow hundreds of times in size, but we go through three distinct stages of development. The first as I have just mentioned, is the fertilised egg. Then around a fortnight after conception, the egg begins to repeatedly divide itself to become a mass of cells, known as the embryo. Two months later, this embryo has grown to approximately 2cm long and is referred to as a foetus.

Extension: *Singing 'the song'*

Aim: *To improve the intonation and rhythm of students' speech.*

Preparation

- Students will practise using the listening script on page 83.

Procedure

- Drill each sentence or clause of the speech using the recording as a model. First drill each line chorally, then one or two individual students.
- When you have done the whole speech, or a good part of it, ask students to practise it to themselves at their own speed. Then ask them to read out the speech to their partner.
- Finally, invite one or two volunteers to read it out to the class.

B

- Ask students to skim through this summary of the second part of the talk. Then ask them to focus on each question. For each question, they should identify the question being asked. For example, for Question 1: *What is the 'baby' at conception?* identify the part of speech (noun) and identify the type of answer required (an object). Ask students to work in pairs to analyse the remaining questions in a similar way. Check answers as a class.
- Ask students to work in pairs to answer questions.
- Students should then complete the summary with information from the speech bubble in Section A, using no more than three words for each answer.

ANSWER KEY

Question 1: 'What is the baby at conception?' (noun; object)

Question 2: 'How many developmental stages does the baby have?' (noun; number)

Question 3: 'What does it divide into after two weeks?' (noun; object)

a Omitted information: comments such as 'quite amazing, isn't it?', 'As I have just mentioned', 'Not only do we grow hundreds of times in size'.

b Synonyms/paraphrase: at the time of our conception: at conception; in the following nine months before our birth: whilst inside the mother; around a fortnight after conception: after two weeks; two months later: after two more months

The answers are: 1 egg; **2** three; **3** mass of cells

C

- Tell students they are now going to explore the relationship between the answers to Questions 1–3 and features of speech. Ask students to listen to the recording 8.2 one last time, reading the summary as they do so. After they listen, they should discuss the questions. After a minute or so, elicit some feedback from the class. You should steer the discussion towards the following conclusions: the information you need to complete the summary tends to be stressed and delivered at a slower pace, often with pauses either side of it. (You should warn students that while this is generally true, they should be careful as in the exam distractors are stressed too.)

D

- Tell students they will now hear the third and final part of the talk, and ask them to complete the summary on the following page, using no more than two words for each question.
- First, ask students to skim through the summary. Then ask them to analyse Questions 4–6 in terms of identifying the question, part of speech and type of answer required. Remind students to try to predict any answers they can, and to listen carefully for features of speech, using them as guides to the correct answer.
- Play recording 8.3.
- Check answers as a class.

Support

- With less proficient classes, you could play the recording again, but remind students that they will hear it only once in the exam.

ANSWER KEY
4 8cm; **5** baby; **6** cry

8.3 LISTENING SCRIPT

Around the end of the first trimester, the foetus has grown to approximately eight centimetres in length and is beginning to resemble a miniature baby, complete with upper and lower limbs. After a further four months, in case you've lost track of where we are, that's seven months after conception, the foetus is approximately 40cm in length. What's more, by now it has a fully developed reflex system, giving the foetus the capacity to breathe, swallow and even cry. This is the reason why, if the mother gives birth prematurely – before the baby's full nine months is up – there is a good chance that it will survive and go on to lead a normal, healthy life, though it might need to live in an incubator for a few weeks.

E

- Tell students that while most multiple-choice questions in IELTS require only a choice of one answer from three options, students may sometimes have to choose more than one answer from a range of possible options. These questions should be approached in much the same way as standard multiple-choice questions: identify the keyword(s) in the stem and each of the options; consider any synonyms for these keywords; try to predict what the answer may be; listen out for the keywords or any synonyms; be careful that you choose the correct answer and not a distractor (see Unit 2).
- Remind students to check how many answers they should choose.
- Play recording 8.4 and ask students to answer Questions 1–4.

Support

- With less proficient classes, you could play the recording a second time, but remind students that they will hear it only once in the exam. Elicit answers from class. If any students have selected any of the distractors, you could ask them to check with the listening script on page 128. Ask if awareness of features of speech helped with this task-type.

express tip

- Read out the advice in the box while students follow in their books.

ANSWER KEY
B, D, E, F (in any order)

8.4 LISTENING SCRIPT

When a baby is born, it can do only a fraction of what it can do in later life. Unlike other species – horses for example, which can walk within a few hours of being born – human babies take several months to learn how to crawl and several more before they can walk. Though fairly helpless, a newborn baby can see. And while its vision is not that developed when first born, it can usually make out faces up to 30cm away. A baby's ability to identify people is not just restricted to sight. Smell plays a part too and newborn babies can identify particular smells – its mother, for instance, and will smile on her approach. Many babies suck their thumbs. In fact, scans have revealed that some do this before birth. They can also swim. While this skill is quickly lost and has to be relearned in later life, when placed in water, babies will do a sort of doggie paddle, swimming a bit like a dog. Although newborns have a walking reflex – pregnant mothers often report feeling the baby 'kick' – what they can't do is walk, or even crawl or turn over. Movement of the head is restricted, too. The lifting of the head from a prone position cannot normally be achieved until around one month, though turning the head is within the range of most newborns.

4 Short-answer questions

Aim: To introduce and give practice in short-answer questions.

- Tell students that they will hear a talk about factors governing our ability to learn how to walk. The talk will be related to three IELTS task-types: short-answer questions; multiple-choice questions with multiple answers; summary completion questions.

Challenge

- For the purpose of this exam practice, the speech has been broken into three distinct sections, each relating to a different task-type, each with a *for this task* box to be read through before attempting that task-type. However, with more competent groups, and to mirror the exam more authentically, you might choose to look at all *for this task* boxes first, then play recordings 8.5–8.7 all the way through.

for this task

- Read out the information in this box while students follow in their books. Elicit answers to questions in the box, for example: *How many words should be used for each answer?*
- Direct students' attention to Questions 1–3. Point out that it is not so much a question as an instruction. Ask them what question is being asked here. (Which three parts ...?) Ask them to identify the keywords in the question (three; parts; crucial; neuro-muscular maturation).

EXAM PRACTICE

Questions 1–3

- Tell students they will now hear the first part of the talk. They should try to follow the instructions in the *for this task* box and answer Questions 1–3.
- Play recording 8.5.
- Students compare answers with their partners.
- Go through the answers now, or check all the answers together at the end of the talk.

ANSWER KEY
brain, muscles, skeleton (in any order)

8.5 LISTENING SCRIPT

Early pioneers described the development of an infant's motor skills in great detail. In the 1930s and 40s, Arnold Gesell identified 22 stages in the development of crawling, beginning with the lifting of the head from a prone position and ending with an even, balanced crawl on hands and feet. Myrtle McGraw, similarly identified seven primary stages in the development of walking, from a newborn's stepping movements to the baby's ability to walk independently by the end of its first year. For these pioneers, motor development was a consequence of neuromuscular maturation, that is mainly independent changes in an infant's brain, its muscles and last but not least its growing skeleton.

5 Multiple-choice with multiple answers

Aim: To practise multiple-choice with multiple answers.

for this task

- Before you play the second part of the talk (recording 8.6), ask students to read through the information in the box.

EXAM PRACTICE

Questions 4–7

- Now direct students' attention to multiple-choice Questions 4–7, and ask them to work through the *Before you listen* stage in the *for this task* box.
- After 30 seconds or so, play recording 8.6.
- Allow students to compare answers with a partner.
- Again, you can give the answers now or wait until they have heard the third and final part of the talk.

express tip

- Read out the advice in the box while students follow in their books. Point out to students that writing just the code will save them a lot of time in the exam and will also save them from having to worry about spelling mistakes.

ANSWER KEY
4–7 C, D, E ,F (in any order)

8.6 LISTENING SCRIPT

This theory of neuromuscular maturation became the popularly accepted explanation for motor development for the next forty years or so. It was not until the 1980s that new research methods and technologies allowed researchers to analyse and measure the development of infants' motor skills in a different way. One such way is the Dynamic Systems Approach, which was developed by the psychologist, Esther Thelen, building on the work of a Russian physiologist, Nicholai Bernstein. In this account, new motor skills are believed to emerge from the coming together of a variety of interacting factors. For example, in order for a child to walk independently, a number of these factors must be in place: the child's muscles must be powerful enough to counteract the effects of gravity. As mentioned earlier, the stepping instinct is common in newborn babies, but they lack the bodily strength to maintain an upright position. However, when they are placed in water, thus making them lighter, they begin to make stepping motions. When they are removed from the water, the action ceases. The stepping reflex normally disappears after a few months. By the way, as I'm sure many of you will know, newborn babies can swim, however this ability is lost with age and has to be relearned. In order to walk, a child should also have lost the top-heavy body proportions typical of infants. The resulting lowering of its centre of gravity gives it better balance and means that it does not have to hold on to things in order to remain upright. They also need a reason to walk. If the baby has no need to go anywhere, why should it? Very young babies cannot see that well, but as its vision and brain matures, it can identify objects from a distance and so its interest is aroused. At the same time, this improvement in perception makes it more aware of its environment. In other words, it can identify the nature of its surroundings and the type of terrain it needs to traverse, making progress possible.

6 Summary completion

Aim: To practise summary completion questions.

for this task

- Ask students to read the *for this task* box. Ask comprehension questions of some key points: *Can you use words in a different form from those in the recording?* (no) *Why should you note your answer as you listen?* (answers often come close together)

EXAM PRACTICE

Questions 8–10

- Now ask students to look at the summary and prepare to answer Questions 8–10. After 30 seconds or so, play recording 8.7.
- Ask students to check answers with a partner. Elicit feedback and check the answers as a class.

ANSWER KEY

8 world; **9** the environment; **10** body parts

8.7 LISTENING SCRIPT

Perception plays a more important role in another approach to motor skills development – the Perception-Action approach, which was inspired by the work of Jane and Eleanor Gibson. For them, there is a strong correlation between our perception of the world around us and our ability to perform movement within it. In other words, our ability to move is not just down to the physical development of our bodies, but also our perceptual ability. For an action to be planned and executed successfully, we need to have perceptual information about certain properties of the environment, our bodies and the relationship between the two. At the same time, we usually acquire sensory information through the use of movement. For example, we may use exploratory movement of body parts such as the hands, feet, eyes and head, to generate perceptual information. In a similar way, actions generate more information for perceptual systems. Furthermore, motor development does not stop after infancy. After mastering basic postural, manipulative and locomotor skills, children acquire many more abilities: writing, playing an instrument, etc. While movement is stiff, wasteful and uncoordinated at first, with practice it becomes progressively more rhythmical, smooth and efficient.

Extension: *Note taking and summarising*

Aim: *To give students practice in listening and taking notes during extended academic talks or lectures.*

Preparation

- You will need two CD/cassette players for this activity.
- Prepare two talks, each 5–10 minutes long, for the class to listen to. This could be something that you have prepared yourself, or perhaps a talk prepared and delivered by a member of the class. Alternatively, use an extract from a television documentary or extended monologue available on the Internet. The BBC website (www.bbc.co.uk) has a wide variety of material in video or audio form. The key thing is that the talk should be fairly academic in subject and tone: the BBC's history or science pages are ideal.

Procedure

- Divide the class into two groups and provide each group with a CD/cassette player.
- Play the talks for each group.
- Ask students to take notes on the speaker's key points as they listen. Encourage them to take notes, rather than write whole sentences. When they have finished, let them check with a partner from the same group. Provide any vocabulary students might need.
- Have students summarise their notes within their groups and provide a summary of the talk to the other group. Allow students to ask questions, then conduct class feedback.

WRITING

Section aims:

- To teach students to recognise which approach to take for Task 2 questions: argument-led or thesis-led.
- To teach students how to write an argument-led essay for Academic Writing Task 2.

1 Introduction

Aim: To generate discussion and elicit vocabulary on some of the themes presented in the unit.

A

- Direct students' attention to the photos and ask them questions such as: *What do the photos say about family relationships? Do you think they are realistic photos? What do families need to do to stay happy? Do you think that having younger parents makes it easier to communicate with them?*
- Ask students to answer Question A.

ANSWER KEY
babyhood; childhood; adulthood; old age

B

- Ask students to look at Question B, and check they understand the vocabulary listed for the advantages of middle age.
- Put students into pairs or small groups to discuss the disadvantages. Go around the class and help students with vocabulary, asking prompt questions to elicit ideas.
- Ask students to read out their answers, and write them up on the board to provide the correct spelling for students to copy down in their books.

ANSWER KEY
Suggested answers:
worries about getting old; being overweight; starting to go bald (in men); heavy financial burdens (mortgage, etc)

C

- Put students into pairs or small groups to brainstorm the advantages and disadvantages of being a teenager. Ask students to read out their answers and encourage students to record these in their books for future reference.

Challenge

- For students who complete this activity quickly, have them brainstorm the advantages and disadvantages for other age-groups (babyhood, childhood, adulthood, old age).

IN THE EXAM

- Direct students' attention to the *IN THE EXAM* box. Ask students to read through the information in the box about the Task 2 Writing exam. When they have finished, have them close their books and elicit the differences between Part 1 and Part 2 of the Writing exam.

2 Deciding the approach

Aims: To expose students to the most frequently appearing exam questions.

To teach students when to use a thesis-led approach and when to use an argument-led approach in their writing.

To teach students how to write an opening paragraph using either approach.

A

- Explain the aims of the task to the students and have them read through the four exam questions and decide which ones take a thesis-led approach (looking at one side of the argument) and which ones take an argument-led approach (looking at both sides of the argument). Students should already be familiar with the thesis-led approach, which they encountered in Unit 4.
- Ask students to compare their answers and discuss how they arrived at each answer. When checking answers, talk students through how a typical IELTS Task 2 question is constructed: first there is a discursive statement, which can include either one point of view (as in Questions 1 and 3), or two opposing points of view (as in Question 2). It can also include a factual statement, as in Question 4. It is the second part of the question rubric, however, that tells us whether the candidate must discuss *both* sides of the argument, i.e. use an argument-led essay (such as Questions 2 and 4). This second part of the question can be phrased as a question (as in Questions 1, 3 and 4) or as a statement (as in Question 2).
- When discussing answers, remind students that argument-led essays are appropriate for ALL types of IELTS Task 2 questions as they are ALL discursive questions, which means students can discuss both sides of the argument for any question. However, there are

some questions (Questions 1 and 3) where students have the choice of simply giving their opinion on the basis of what they think and feel (thesis-led approach).

Answer key

thesis-led: 1 and 3

argument-led: 1, 2, 3, 4

express tip

- Read out the advice in the box while students follow in their books. This box provides further information regarding the differences between the two types of essays. This information will help them answer Question B1.

B

- Ask students to read through the two opening paragraphs and have them answer Question 1.

Support

- If students find this difficult, remind them of the *express tip* box and ask them the following questions: *Which opening paragraph includes the personal opinion of the writer?* (**a**) *What language is used to indicate personal opinion? (In my view) Which opening paragraph presents a balanced view of the topic?* (**b**).
- Ask students to answer Question 2 by matching the opening paragraphs to the appropriate exam question in 2A.
- Check the answers as a class, making sure students justify their answers. For example, both Question 1 and 4 deal with the generation gap, so how did students decide which is the correct answer?
- Question 3 provides more language students can use in an opening paragraph, depending on the approach they take (thesis- or argument-led). If students have difficulty here, remind them again of the information in the *express tip* box, i.e. argument-led essays do NOT contain expressions of personal opinion, thesis-led essays do.
- Students should now be in a position to answer Question 4 and write their own opening paragraph using the argument-led approach (Students practised the thesis-led approach in Unit 4). Encourage students to use the model to structure their answer. You could also point out here the use of the rhetorical question, which is common to argument-led approaches. Note that students are expected to use appropriate language (as seen in Question 3) and to refrain from using personal opinion: for argument-led essays, the personal opinion of the writer is left to the end of essay.

Answer key

1 a is thesis-led; **b** is argument-led

2 a matches Question 3; **b** matches Question 4

3 Argument-led essay: Many people think, It is generally considered that

Thesis-led essay: I believe, In my opinion, I totally disagree with

An opening paragraph for a thesis-led essay states the personal opinion of the writer from the very beginning and deals exclusively with this side of the argument. For an argument-led approach, the words in the exam question are presented again to introduce two sides of the argument without presenting an opinion at this stage.

C

- This task can be set for homework if class time is short.

Answer key

Suggested answer:

Many people feel that the gap in understanding between the older and younger generation is too wide to be bridged. In the first case, the question assumes that there always is a 'gap' and that it needs to be bridged. In the second case, we need to look more specifically at what is meant by 'older' and 'younger' people. Are we talking about differences between teenagers and middle-aged people, or differences between young adults and middle-aged people?

Extension: *The debate*

Aim: *To practise brainstorming ideas 'for and against'.*

Preparation

- For a class of students who generally have a lot to say, set up a debate to answer essay Question 4 from 2A.

Procedure

- Direct students to the rubric in Question 4 from Section 2A.
- Divide the class into two groups and give each group 'a point of view' to prepare: Group A looks at the case 'for', Group B the case 'against'. After a few minutes' preparation, ask for a volunteer speaker from each side to put forward the case. Ask for a second speaker from both sides to support the first speaker, bringing up any points not previously mentioned, emphasising and further illustrating some of the key points mentioned.
- The final part of the debate can be opened up for questions 'from the floor'. Finish off with 'a show of hands' vote.

3 Providing supporting evidence

Aims: To teach students how to write and structure the main body of an argument-led essay.

To teach students how to support the main idea of each paragraph with supporting evidence.

A

- Ask students to read through this first paragraph of the main body of the essay and match the paragraph to the correct exam questions in 2A. Ask students to check their answer in pairs, then as a class.

ANSWER KEY
Question 1
Note Students should see pretty quickly that it is about the 'generation gap' and that the answer can be either exam Question 1 or 4. As there is no specific reference to 'exchanging daily routines', students should conclude that this main body paragraph is talking generally about 'generation gaps' and therefore the answer must be exam Question 1.

B

- Direct students' attention to the words in the language box and check they understand them. Explain that these phrases are used for presenting supporting evidence.
- In pairs, have students complete the gaps in the paragraph on the previous page, using the phrases in the box. Check answers as a class.

Support

- Vocabulary checking questions you could use are: *What two expressions have a similar meaning?* (*for instance, a good example of this*); *Which expression means the same as 'emphasise'?* (*highlights the point*); *Which expression indicates that contrasting information will follow?* (*but this is not the case for*); *Which expression indicates what the writer has actually done or seen something herself?* (*from my experience*).

ANSWER KEY
1 A good example of this is; **2** for instance; **3** but this is not the case; **4** From my experience; **5** highlights the point

C

- Explain that a well-written paragraph should always contain one main idea – when we change ideas, we change paragraph. Students will have done a lot of work on deconstructing paragraphs in Unit 1 (topic sentences, paragraph summaries, etc), and should be aware of this, but it is a good idea to remind them and make the connection explicit between reading and writing skills.
- Ask students to read the paragraph to note down the main idea and supporting information. Ask students to complete this activity in pairs, then check answers as a class.

ANSWER KEY
Main idea: different generations have different experiences = generation gap
Supporting info: technology used differently by different generations; young people are brought up with technology — older people were not; personal example of parents not knowing how to program a VCR and slow at texting

D

- Ask students to read through the notes and write their own paragraph based on the notes.
- Make sure students understand that this second main body paragraph follows on from the previous one in so far as it is another paragraph arguing that the generation gap is too wide. Remind students to use appropriate language to set out supporting information as introduced in 3B.

ANSWER KEY
Suggested Answer:
Another main reason why some people think the generation gap is unbridgeable is because both older people and younger people like it that way. Naturally many older people are in positions of authority as parents or bosses and think that if they keep a distance from their children or younger employees they will be respected by them. Younger people, in the same vein, aim to keep a distance from their parents or bosses because they don't want them to know all the details of their lives, as they fear they will be judged and punished if they do.

E

- In pairs, ask students to match the four elements of supporting information (a–d) to the relevant main idea (1, 2).
- Check answers as a class.

ANSWER KEY
1 b, d; **2** a, c

Challenge

- Ask students to brainstorm further supporting information for each of these main ideas.
- Are there any other main ideas and supporting information that can be used here to support the view that the generation gap is not too wide to be bridged?

F
- Finally for Question F, students will write the last two paragraphs for the main body of this essay, using the notes from Section E. Have them start their writing using the words *On the other hand ...*
- As a final check to make sure the students know what they are doing, ask them: *Why are we starting the next two paragraphs with the words 'On the other hand'?* (These two paragraphs present the opposing view, i.e. the 'generation gap' is NOT too wide to be bridged.).

ANSWER KEY

On the other hand, many people would argue that the generation gap is not too wide to be bridged as nowadays many parents and teenagers share similar interests and attitudes. This is particularly the case for parents born after the 1960s which was a period of great social change. Teenagers, particularly late teenagers from 16–19 years old can be said to be much more mature than in the past as the job market today is a lot more competitive, meaning they are less interested in rebelling and more interested in getting a good degree and a good career. In terms of shared interests, this can be seen every Saturday afternoon across the country where fathers and sons both enjoy going to football matches together.

Parents don't only share attitudes and interests with their children, they are also much more likely to understand them better. Parents after all were once teenagers too and experienced many of the same problems and difficulties that their children go through. Parents are also much better able to make their own experiences relevant to today, as in general, they are much better informed on teenage issues through TV programmes and the Internet.

4 Academic Writing Task 2: Essay

Aims: To give students the opportunity to consolidate the skills introduced in this unit.

To practise a complete Task 2 question.

To practise writing within a specified time limit.

for this task

- Read through the information in the box as a class, as preparation for the exam question. You can do this in one go or you can take students through it point by point and conduct class feedback afterwards.
- Ask students to decide what approach they need to take (argument-led). Then have students underline the keywords in the exam question (*conflict between teenagers and parents; necessary part of growing up; something negative which should be avoided*).
- Have students write a brief paragraph plan and put the students into pairs or groups to brainstorm ideas for the two different views.

express tip

- Read out the advice in the box while students follow in their books.

EXAM PRACTICE

- You could continue to go through the remaining points one by one, writing the essay in stages and checking as you go, or you could have the students write the whole essay after having done the plan. The former option might be appropriate for a weak class, but this option would take considerable time to complete (allow 60–90 minutes). Alternatively you can set this essay for homework.
- When giving feedback to students on their essays, direct them to the model essay on page 109 of the *Coursebook*. Explain to students that the model essay is not a definitive answer – it is just one way of answering the question.

Extension: *Mind-mapping an essay plan*

Aims: *To teach students how to use mind-maps to plan an essay by organising their thoughts and ideas.*

Preparation

- Draw the following mind-map on the blackboard, or photocopy it and provide copies to each student or pair of students.

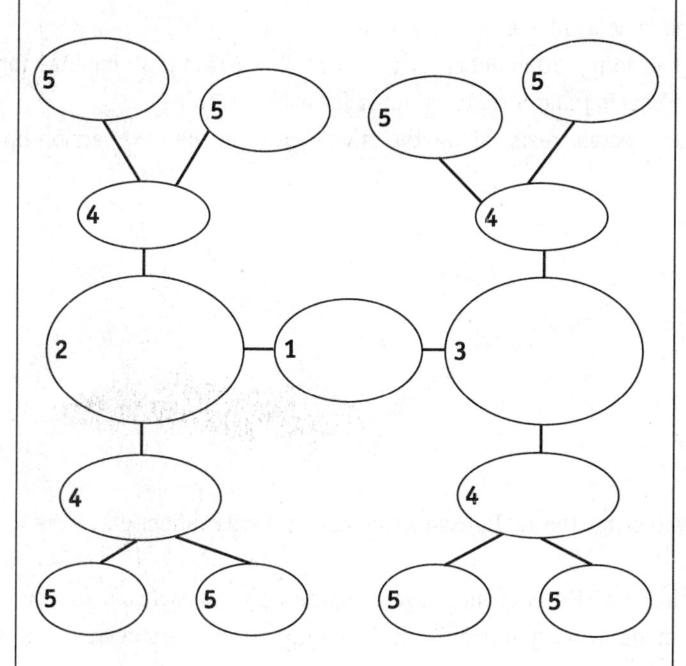

Procedure

- Ask students to work in pairs or small groups. Ask students to match the following labels to 1–5 in the diagram:

a supporting information
b main ideas
c exam question
d view that says conflict between parents and teenagers is a necessary part of growing up
e opposing view that says conflict is negative and should be avoided

(**Answers: 1** c; **2** d; **3** e; **4** b ; **5** a)

- Now ask students to draw the mind-map in their notebooks and brainstorm ideas in pairs or small groups for one of the exam questions contained in this unit. Students should write down their ideas in note form within the circles of the mind-map.

Model answer

We all want to grow up into mature and confident adults, who are emotionally and psychologically balanced. The question is - how can this be achieved? Do we encourage young people to rebel against their parents as a way of 'standing on their own two feet' and establishing their independence, or do we take the opposite view and discourage generational conflict between parents and children by helping both age groups to better understand each other?

Adolescence is traditionally a difficult age when young people go through dramatic physical and emotional changes in a relatively short space of time. As a result, many people feel that some conflict between parents and their children is inevitable. In a sense, it is neither 'good' nor 'bad'; it is just a fact of life. I think it should also be noted that 'conflict' is seen differently by different cultures - in the West it can often be seen in a positive light, whilst in many Asian cultures it has negative associations.

In my own personal case, I didn't communicate very much with my parents while I was growing up and therefore there was no real conflict involved in our relationship, but on the other hand there was no real understanding either. This was a similar experience of many of my male friends. Girls, on the other hand, seemed to have a closer relationship with their mothers, especially the ones with younger mothers, which meant that the age gap was not so great.

Personally, I think that in an ideal world, children should communicate closely with their parents because both could learn a lot from each other. In my view, I think we isolate ourselves too much within the narrow confines of our own age group, giving us a limited view of the world and preventing us from growing up in the best way possible. The problem is, we don't live in an ideal world. Parents and their children are not perfect, so in the end I feel that teenage conflict is probably an inevitable, but certainly not negative, part of growing up.

PROGRESS TESTS

INTRODUCTION

The Progress Tests

IELTS Express Upper Intermediate contains three Progress tests, positioned after Unit 2, Unit 4 and Unit 6. They provide further practice in the skills, task types and exam sections previously covered in the core units, and reflect the actual IELTS exam in terms of content and degree of difficulty. There are no tips to aid the student, nor does the *Coursebook* contain listening scripts or an answer key for these tests. They are, in effect, 'bite-size' versions of the exam. (There is an answer key and listening script in the *Teacher's Guide*.)

It is advisable that students do these tests very soon after completing the preceding skills sections. This is because the Progress tests are designed to consolidate new skills as they are learnt by providing practice in an exam-style context. Each Progress test covers each of the four skills and may be given in its entirety in one session, or each part set individually, depending on how the book is being used and time available.

A note on timing: allow 20 minutes for each Progress test Reading, 20 minutes for each Part One Writing, 40 minutes for each Part Two Writing, 20 minutes for each Progress test Speaking and 5 to 10 minutes for each Listening.

For further information on how to administer and mark the Progress tests, follow the advice given in the next section on the Practice test.

PRACTICE TEST

INTRODUCTION

The Practice Test

The *IELTS Express Coursebook* contains a complete Practice test for the IELTS exam. This test is a true reflection of the different levels of questions and tasks in the IELTS exam.

It is recommended that you use this test as a mock examination after students have completed the *Coursebook*, shortly before they take the actual exam. Answer keys, including model writing answers, and listening scripts are provided in this *Teacher's Guide*.

How to administer the Practice Test

Ideally, in order to give your students as accurate a simulation of the IELTS exam as possible, students should do all of the sections back-to-back in a single day, although sometimes the Speaking exam may be taken on a different day. However, this will require at least $2\frac{3}{4}$ hours for the Listening, Reading and Writing papers, not including short breaks between papers. However, you may prefer to do it a paper at a time, over a few days.

Remember that no dictionaries or any other reference books may be used in any part of the IELTS exam.

Make enough copies of the answer sheet at the back of the *Teacher's Guide* for each student. Tell them that they will have time at the end of each relevant section to transfer their answers onto the answer sheet. Insist that there is no talking throughout the exam and make sure everyone's mobile phone is turned off before the exam begins.

IELTS Part 1: Listening

Time: up to 30 minutes, plus 10 minutes to transfer answers

External noise, traffic, etc can be very distracting during a Listening exam, so try to find a quiet room or a quiet time of day to conduct this section. Students who arrive late will be very distracting to the rest of the class, so strongly encourage students to arrive on time.

You can play the tape or CD on a player positioned at the front of the room, but make sure it is loud enough for everyone to hear. Alternatively, you could use a language lab, but remind students that in the actual test a cassette player will be used.

- Distribute answer sheets to the class and have students write their names at the top of the sheet. Tell students that during the Listening, they are to record their answers in their *Coursebook* and that at the end of the exam they will be given ten minutes to transfer their answers onto the answer sheet.
- Have students open their books to page 90. Play the recording ONCE ONLY. Do not stop or pause the recording, but just let it play through to the end.
- When the recording has finished, inform students that they now have ten minutes to transfer their answers to the answer sheet, using a PENCIL. Tell them they must be very careful to record their answer in the appropriate place – candidates often lose marks for not doing this. Remind them that correct spelling is important and also that they must observe any word limits for particular questions. After five and nine minutes, announce how long they have left.
- Collect answer sheets for marking after ten minutes.

IELTS Part 2: Reading

Time: 1 hour

- Distribute answer sheets to the class and have students write their names at the top of the sheet. Tell students that they are to record their answers on the answer sheet IN PENCIL as they do the exam. Unlike the Listening exam, there is no extra time given for this.
- Ask your students to open their *Coursebooks* at page 94. Explain that students have only one hour to read all three texts. Write the start time and finish time on a board at the front of the room, but remind them that the questions can be done in any order.
- Announce when there are only five minutes remaining. When one hour is up, insist that they finish immediately, then collect answer sheets for marking.

IELTS Part 3: Writing

Time: 1 hour

- Make sure that students have enough writing paper, and have some spare sheets at the front. Have students write their names at the top of each piece of paper they use. Have students use a different sheet of paper for Tasks 1 and 2 – this reflects exam procedure.
- Ask your students to open their *Coursebooks* at page 103. Explain that students have only one hour to answer both questions. Write the start time and finish time on the board at the front of the room. Remind them that they should spend approximately 20 minutes on Part 1 and 40 minutes on Part 2. Tell them they can answer the questions in any order.
- Announce when there are five minutes remaining. When one hour is up, insist that they finish immediately, then collect answer sheets for marking.

IELTS Part 4: Speaking

Time: approximately 1 hour

Each interview typically lasts between 12 and 15 minutes. This means that even with a relatively small class, conducting the practice test individually would take several hours. You may choose to do this, but the procedure outlined below will enable you to save a great deal of time.

Note: If you have the *IELTS Express Speaking DVD*, you should show it to your class before conducting this test to help them understand what they should be focusing on in the exam.

- Make a photocopy of the topic card for Part 2 of the Speaking exam from page 105 of the *Coursebook*.
- Remind the class of the examiner's criteria for the Speaking exam by writing these terms on the board:
  Fluency and coherence; Lexical resource; Grammatical range and accuracy; Pronunciation
  These criteria are explained in the section 'What are the examiners looking for?' in the introduction on page 4 of this book, and also in the *IELTS Express Speaking DVD*.

- Divide the class into four groups and allocate one criterion to each group. As you interview the model candidate, ask the class to listen carefully and evaluate the candidate on their criterion, taking notes on *Good points* and *Areas for improvement*.
- Ask a strong and confident member of the class to do the exam in front of the class. Tell them that you will play the examiner and they are to imagine that they do not know you and you do not know them. Using the questions and question cards on page 105, you will conduct a mock Speaking exam with a student to provide a model for the rest of the class.
- When the Speaking exam finishes, praise the candidate for doing it in front of their classmates. Ask the class for positive comments on each of the examiners' criteria and then areas in which the candidate could improve.
- Now divide the class into pairs or groups of three. The pairs will take it in turns to play examiner and candidate, using the same material from page 105 as the candidate. In the groups of three, one student could act as an observer. Each interview should last at least 10 minutes. Encourage them to give feedback on each other's performance. If you are concerned that students may become familiar with the material if they all use the same questions, you can photocopy the Speaking Topic Cards on pages 123–124 of this book and give one pile to each group. 'Examiners' can then draw one card from this pile at random for the test.
- While the interviews are underway, circulate and monitor taking notes – making sure that you, and the students acting as examiners, are evaluating and giving feedback based on the IELTS criteria. When all groups have finished, invite some of them to comment on each other's performances, and feed in your comments as appropriate.

Marking and grading the papers

Listening and Reading

Use the answer key provided in the *Teacher's Guide* to mark the answer sheets. Answers are considered either right or wrong. There are no half marks awarded. Answers with spelling mistakes are considered as incorrect.

The way in which a candidate's marks on the IELTS exam are converted into an IELTS band score is confidential and varies slightly from one test paper to another. However, an approximate guide to what candidates may achieve can be determined according to the following table.

Note: This chart is approximate and should serve as a guide only. It is by no means definitive.

Band	Listening score	Academic Training Reading score	General Training Reading score
4	9	8	15
5	16	15	23
6	23	23	30
7	30	30	34
8	35	35	37

Writing and Speaking

Grading candidates' performance in these two areas is much more complicated and examiners undergo extensive training to learn how to do this. The examiners criteria for Writing and Speaking on page 4, should assist you in evaluating your students' work. Model writing answers are provided on pages 97–98. For further information visit www.ielts.org.

Progress Test 1 Answer key

Listening

1 yellow; **2** poetry; **3** 7:30; **4** Jazz; **5** Spanish; **6** Simon Evans; **7** 83a York Way; **8** 93002344; **9** C; 10 B

Reading

1 v; **2** x; **3** xi; **4** ix; **5** vi; **6** vii; **7** ii; **8** i; **9** job; **10** air; **11** similarity; **12** addition; **13** money; **14** vulcanisation; **15** (the) price soared; **16** Ceylon (and) Malaysia

Academic writing

Sample answer

Writing Task 1

The table below shows how companies divide their advertising budget between a variety of promotion channels in three different countries.

Overall, all three countries are similar in the sense that the top three promotion channels are the same - TV is the most popular followed by newspapers and radio. However, there is a very sharp contrast between Australia and Korea on the one hand and Brazil on the other. With the former two countries, there is a broadly similar pattern of expenditure which doesn't show that much difference in expenditure between the three channels. Brazil however, has a very unbalanced distribution of expenditure with over 60% of all money spent on advertising in the country done through a single channel - TV.

In terms of the three less popular promotion channels, there is very little similarity between the countries. Australian advertisers obviously feel that the Internet is a very effective channel as they are spending nearly as much here as the combined expenditure on direct mail and sponsorship. Korea is the opposite, spending twice as much on sponsorship and direct mail than on Internet advertising. Brazil, again, is the odd one out, spending very little on these secondary promotion channels.

Progress Test 2 Answer key

Listening

1 G; **2** E; **3** A; **4** F; **5** peas and beans; **6** high nitrogen requirement; **7** spread of weeds; **8** (removable) wooden; **9** brick base; **10** (old) carpet

Reading

1 secret entrance; **2** three granite rocks; **3** Grand Gallery; **4** (the) King's Chamber; **5** (the) Queen's Chamber; **6** C; **7** A; **8** B; **9** C; **10** D; **11** highly polished limestone; **12** (the) north face; **13** vermin

Academic writing

Sample answer

Writing Task 2

I partly agree and partly disagree with the statement in the question. Some people such as business people do earn a lot of money and are incredibly busy, whilst other people are unemployed and receive no money at all. There is however, a group of people in the middle such as teachers, a lot of office workers as well as all part-time staff who work less hours and receive less income. The question is - is society as a whole unbalanced?

I think the answer to this question depends on the society you are talking about. Some countries in the world such as countries in Western Europe have a more equal distribution of income and have laws which limit the number of hours you can work in a week. They also have a high number of public holidays and guarantee workers a minimum of 20 days holiday per year. Other countries have no such laws. In the USA, the standard work contract includes only 10 days holiday per year. In Japan there is a also very strong work culture which means that many workers do not take their annual holidays and do a lot of overtime during the week.

In many developing countries, there is not enough work, but that is not to say that they have 'time on their hands'. From my experience, most people without a job in developing countries spend their time looking for work in the cities or working on their own small plot of land in the country. They have very little time and very little money.

Is there much difference with countries with highly developed economies? Not much in my view. The cost of living is very high in industrialised countries, so people working in these 'rich countries' might earn a lot of money but they also need to spend a lot of money to survive.

To sum up, I would say that we do live in a world that is unbalanced, but it is not so much an imbalance between those who have time and those who haven't. The main imbalance in my mind is between those people who have money and those who do not. Many 'seriously rich' people, who have inherited their wealth have lots of money and lots of free time!

Progress Test 3 Answer key

Listening

1 (agro-)chemicals; **2** crop sowing; **3** grasslands; **4** E; **5** C; **6** F; **7** loss of biodiversity; **8** (agricultural) subsidies; **9** nesting and feeding; **10** a/one fifth

Reading

1 Not given; **2** Not given; **3** Yes; **4** Yes; **5** No; **6** Yes; **7** F; **8** A; **9** G; **10** C; **11** early space rocket; **12** roller blinds; **13** windows; **14** underground entrance

Academic writing

Sample answer

Writing Task 1

The graph illustrates how the volume of traffic increases and decreases over a 24-hour period from midnight to midnight in the London area. The graph not only shows the total volume of traffic, it also shows how this is broken down in terms of total passenger numbers for bus, car and underground.

The main thing to note from the graph is that there are two clear rush-hour periods in the morning between 8am and 10am and in the evening between 6pm-8pm where traffic levels are very high. There is also a third rush-hour period in the afternoon from 2pm to 4pm, where traffic levels are relatively high compared to the rest of the day. This afternoon rush-hour is not as busy as the morning and evening rush hours, with just over three million people travelling, compared to well over six million in the morning and evening peak periods. It is important to point out that the key factor in this afternoon rush-hour is car traffic, which jumps dramatically over this two-hour period, contrasting with bus and underground passenger numbers, which remain relatively constant.

In terms of total numbers of people travelling, we can see that the underground is by far the most popular form of transport for Londoners, although it is clear from the graph that the system does not run 24 hours a day as it looks to be closed from 3am-5am. The bus is the second most popular form of transport but unlike the underground the total level of passenger numbers fluctuates much less. The car, on the other hand, is surprisingly little used in London with the exception of the afternoon rush-hour - perhaps this is a result of high levels of congestion and expensive parking.

Listening

Section 1

1 Hornby; **2** Ilford; **3** 94456781; **4** coffee table; **5** 39.99; **6** handbag; **7** DBR29; **8** Express; **9** A,D; **10** D,E

Section 2

11 clothing; **12** determination; **13** relaxation; **14** diet; **15** body care; **16** B; **17** C; **18** A; **19** B; **20** B

Section 3

21 specialist magazines; **22** studio websites; **23** (local) reviews; **24** A; **25** B; **26** income; **27** reactions; **28** consistent; **29** Re-shoot; **30** genre

Section 4

31 eastern; **32** piracy; **33** repairing; **34** slaves; **35** (large) houses; **36** fish sauce; **37** specialised; **38** B; **39** C; **40** F

Academic reading

Reading passage 1

1 (lozenge-shaped) washers; **2** (tiny) dots; **3** subsoiling machine; **4** excavating spade; **5** potato field; **6** a tin worker; **7** the Ringlemere Cup; **8** treasure; **9** (the) farmer; **10** ritual landscape; **11** B; **12** A; **13** D

Reading passage 2

14 Yes; **15** No; **16** Not given; **17** No; **18** Yes; **19** Yes; **20** Not given; **21** Yes; **22** E; **23** C; **24** G; **25** A; **26** C

Reading passage 3

27 Not given; **28** Not given; **29** False; **30** True; **31** cognitive therapy; **32** six/6 months; **33** personality traits; **34** health anxiety; **35** C; **36** E; **37** B; **38** A; **39** D; **40** C

Academic writing

Sample answers

Writing Task 1

The charts show that there were big increases in the sales of the larger vehicles in some months, when compared with sales of these vehicles in the same months of the previous year. These increases were greater for large SUVs than for large pick-ups.

Both large SUVs and large pick-ups had lower sales in the first three months of the year in comparison with the same months the previous year. However, apart from two months for large SUVs, sales were higher both for large SUVs and large pick-ups in all the other months when compared with the previous year.

The sales of other vehicles did not vary so greatly. They decreased in more months than they increased in comparison with the previous year, but falls and rises were smaller than for the larger vehicles.

The overall picture is that, despite their high consumption of fuel, there were some big increases in the sales of new large vehicles.

Writing Task 2

Although it is true that a great many people now have a higher standard of living than in the past, it does not seem that this necessarily makes people any happier. In many ways, and in many places, life is much easier and more comfortable than it used to be, but it's possible that in the past people were happier with less.

One feature of modern life that has changed in many places is that people have less contact with each other and there is less of a feeling of belonging to a community. People leave their hometowns more than they used to and so family ties are often not as strong as they used to be. In cities, people often do not know their neighbours. As a result of all this, people can feel isolated. People help each other less and there is much more emphasis on the individual.

Modern life has also brought with it problems that people do not think existed in the past. For example, many people are concerned about rises in crime, and they also feel that life is much faster and more stressful than it used to be.

Another reason for people's unhappiness with modern life is that they have higher expectations than they used to have. They have more but they want even more. This leads to feelings of disappointment and even anger when people do not have everything they think they should have. People take for granted all sorts of comforts and material possessions that people in the past would have considered luxuries. They don't appreciate what they have; they simply want more.

It is also true, in my opinion, that when people have all their basic needs for survival, they are likely to become more introverted and think about life more. This can lead to feelings of depression and dissatisfaction with their lives. If you are struggling to survive, you do not have the time or the opportunity to think so much about the meaning of life.

Of course, it is impossible to return to the past and nobody would want to exchange the comforts of modern life for the way that people used to live. However, in many ways progress has not made people happier, in my view.

Progress Test 1 Listening scripts

Listening

Artslink: Good morning, Oldbridge Artslink.

Caller: Hello. Is this where you get information about the arts festival?

Artslink: That's right. Did you want some tickets?

Caller: Well, maybe. But I wanted to find out what was happening first.

Artslink: Have you seen the programme?

Caller: No, not yet.

Artslink: Well, let me see. Yes. This year we have events at three different venues. The first is The Old Bus Station Arts Centre.

Caller: Ah, yes. Opposite the library.

Artslink: That's right. The second venue is the Town Hall. The third is Lincoln Park. Last year everything was there but it rained so heavily they decided to move some of the events indoors for this year.

Caller: Good idea! So what's on?

Artslink: Well, in the afternoon between two and four, at the Arts Centre, there's a play for children called 'Uncle Spoon and the Yellow Balloon'.

Caller: Right ...

Artslink: They're very good. I took my daughter last year.

Caller: I'll probably give that one a miss.

Artslink: OK. At the same time at the Town Hall, there's 'Chasing Clouds', which says here is poetry readings. Meanwhile, in the park, there's a folk singer, Nick Goose.

Caller: Oh, I think I've heard of him.

Artslink: The next lot of events are on in the late afternoon to early evening, starting at around five and due to finish at half past seven.

Caller: OK, so what's on then?

Artslink: A storytelling workshop in the arts centre. Over at the Town Hall they've got three new short plays. They go under the title 'First Steps'. And in the park, there's a band called 'Trad Dad and the Modern All Stars'.

Caller: What sort of music do they play? Pop? Rock?

Artslink: Er ... No. Jazz, I think.

Caller: Sounds interesting.

Artslink: Then in the evening, you've got 'Gypsy Ballads'; that's Spanish dance at the Arts Centre. Stand up comedian Eddie Hicks is at the Town Hall, supposed to be very funny, and in the park you've got a heavy metal band called 'Maidenhead'.

Caller: Maidenhead?

Artslink: Yes, very loud, they say. You like that sort of thing, do you?

Caller: Well, I don't mind, yeah.

Artslink: So, does any of that interest you?

Caller: Yes. Can I buy a ticket from you?

Artslink: Unfortunately not. We've been having problems with the system this morning. I can send you a ticket application form if you like.

Caller: Yes, please.

Artslink: If I can just take your details. What name is it?

Caller: Simon Evans. That's S-I-M-O-N E-V-A-N-S.

Artslink: OK, and can I take your address, please?

Caller: Certainly. It's 83a York Way. That's Y-O-R-K Way. Number 83a.

Artslink: And that's Oldbridge, is it?

Caller: Yes. Do you need my phone number?

Artslink: Yes, please.

Caller: It's 93002344.

Artslink: And what sort of ticket did you require?

Caller: Is there a choice?

Artslink: Oh, yes. A group ticket is twenty pounds per person.

Caller: How many in a group?

Artslink: Six or more.

Caller: Oh, no, it's just me. Just the one.

Artslink: And you're over 18?

Caller: Yes.

Artslink: Oh, shame. You could have had a child's ticket for twelve pounds. So it looks like you'll have to pay the regular.

Caller: And how much is that?

Artslink: Twenty five pounds. But for that you can see anything you like.

Caller: OK. That's fine.

Artslink: And can I just ask you how you came to hear about our arts festival?

Caller: Er, I do remember seeing a poster at a bus stop. But that was for a circus. Could it have been a newspaper? I don't think anyone told me about it either. No, it must have been in the paper.

Artslink: Fine. I'll get this in the post to you this afternoon and I hope you enjoy the festival.

Caller: Yes. Thank you. I hope so too.

Speaking

Part 1

Examiner: Have you travelled far to come here today?

Candidate: No, not far. I live a couple of miles from here in Insurgentes Sur.

Examiner: How would you describe your local area?

Candidate: It's fine, but there are a lot of busy roads so it does get a bit noisy sometimes.

Examiner: What's the best thing about the area where you live?

Candidate: Well, for me the best thing is having all my friends and family nearby. My cousins live in the same building as me and my best friends live in an apartment block opposite.

Examiner: What's your favourite part of the city?

Candidate: Polanco. There's a great atmosphere there and there's no problem finding places open in the evening to meet up with friends.

Part 2

Well, that's easy! The one possession I could not live without is my mobile phone. I use it all the time – mainly to chat to my friends or text them about where and when we are meeting up. This phone I've got also takes photos, videos and is an MP3 player, so I can use it for all kinds of things. To tell you the truth though, the picture and sound quality are not so good, so I don't use these functions so much.
What would I do if I lost it? Well, I'd probably go out the next day and buy a new one! One good thing would be that I could get the latest model phone where I could connect to the Internet on the move. It would be great! The only problem with losing my phone is that all my friends' numbers are programmed into my address book and so if I lost it, I would be really upset.
Basically, my mobile is so important to me because it keeps me in contact with everyone – my parents can reach me whenever they need to; my friends can call for a chat. I can play games on it. I never leave home without it!

Progress Test 2 Listening scripts

Listening

These days more and more people are taking up growing their own vegetables. Now, getting started growing your own vegetables isn't too tricky if you follow a few basic rules. The first is how to lay out your vegetable garden or allotment. Of course, there are as many different vegetable gardens as there are vegetable gardeners, but most have more or less the same elements in common. So, let's take mine as an example. We've got a little map of it here.

You enter the allotment through a little gate at the bottom here. Immediately on your left is the manure heap and on the right are the all-important <u>compost bins</u>. We'll talk about how to construct these in a minute. Back to the map. Looking straight ahead, you see a pathway which runs through the middle of the <u>vegetable beds</u>, there are two beds on either side of the path.

Carrying on up the path, on the left is the strawberry patch. Try to keep this area free from weeds or pests such as slugs, and you'll be rewarded with delicious succulent fruit. Directly opposite, on the other side of the path, is a bed full of <u>tall plants</u> – sweetcorn, Jerusalem artichokes and the like. These give you a bit of privacy when you're relaxing in the sitting-out area just behind it. Finally, in the top right-hand corner of the plot is the shed, where you can keep all your tools. In the other corner is a <u>greenhouse</u>, perfect for growing chilli peppers, courgettes or even grapes.

Right, let's get back to those four narrow beds. Why have four of them? Well, over the years gardeners have found that using what's called a rotation system gives the best results. In this system, closely-related vegetables are grouped together and are all grown in the same bed for one year and are then all moved to a different bed for the next year. So, let's look at what we'd grow in one bed over a three-year period. In the first year, we'll have legumes, <u>vegetables like peas and beans</u>. Now legumes improve the soil by releasing nitrogen into it, so they are followed in the second year by the brassicas; things like cabbages and broccoli, which have a <u>high nitrogen requirement</u>. In the third year, we'll plant vegetables from the potato group. This includes potatoes, obviously, but also tomatoes, peppers and aubergines. These plants can prevent soil sickness and also, with their large foliage, help to stop the <u>spread of weeds</u>. So, if we've got a three-year rotation cycle, why have we got four beds? Well, the fourth bed is for your permanent crops, such as rhubarb, raspberries and globe artichokes.

Now, as I said earlier, the most important thing in your vegetable garden is probably your compost bin. This provides you with lots of decayed organic matter to dig into the soil, to feed it and create the perfect conditions for growing vegetables. There are many types of compost bins commercially available, but it's much cheaper to make your own. Here's how: first, build a wall out of breeze blocks in the shape of a capital letter 'E'. This will give you two compartments and allow you to use compost from one bin while the second is still being built up. Either side of each opening, place two upright posts. The front of each bin is composed of <u>removable wooden</u> boards which can be slotted behind the posts. This gives you easy access to the compost, but stops it falling out. The composting process will be improved if there is adequate ventilation. So each bin has a <u>brick base</u> for this purpose, on top of which is a rigid mesh screen. Whatever you want to compost – vegetable peelings, tea bags, eggshells, grass cuttings and so on – can all be added to the bin. As the compost decays, it generates heat, which in turn speeds up the process, so it's important to keep the heat in. You can do this by covering each bin with a piece of old carpet. Don't forget to weigh it down with some bricks. Moving on now, let's look at dealing with garden pests ...

Speaking

Part 2

I did a successful presentation for a project last semester in my international marketing class. It was a group presentation using Powerpoint, but we had loads of preparation to do before the presentation. We had to choose a company and then do a lot of research on the Internet to find out more information about it. Our team chose Zara, which is a Spanish clothing company. It was very difficult finding all the information, but I really enjoyed it because I found some very interesting websites.

Anyway, after we had done our research, we had to put it all together in the form of a presentation, which we had to give to the teacher and the rest of the class at the end of the semester. I was really nervous because I hadn't done a presentation before – we had to stand up in front of the class and speak for about five minutes each! Being so nervous was probably quite good for me because I made sure I did a lot of preparation beforehand. Basically, I practised a lot at home in front of the mirror, and we also did two practice runs in our group.

When we did the presentation, I managed to speak without using my notes very much and actually enjoyed being up there. This was the important part for me. I knew it went well because we got some very good feedback from the teacher and the other students in the class – it was a good feeling!

The reason why I think it was successful is because our group worked well as a team and we spent a lot of time practising. Some groups however, spent too much time on the research part of the project and not enough time creating a good presentation – the worst presentations were when people just read from their notes and didn't even look at the audience while speaking.

Part 3

Examiner: How would you define success?

Candidate: That's a difficult question, but basically I think you can define success in many different ways. For me though, the key thing is happiness – if you are happy, you are successful in life.

Examiner: Do you think society places too much emphasis on money as a measurement of success?

Candidate: Yes, I do. There is this idea that money buys you happiness and that if you don't have money you're not happy and not successful. Magazines, TV, posters on the street, all keep telling us to buy more things – it seems like the more things we have, the more successful we are. It's difficult not to be influenced by it.

Examiner: What do you think distinguishes successful people from unsuccessful ones?

Candidate: I don't know. Maybe it's a mix of many things – motivation is important and so is hard work. You need to believe that you can do something and then you need to work really hard to make it happen, I guess.

Examiner: What makes some companies successful and other companies not?

Candidate: I studied business at university – in our course we looked at some successful companies that were really creative. They invented new products that people wanted, and invested a lot of money in good design and cool branding. Unsuccessful companies, I think, don't really understand their customers and don't understand that their needs and wants are always changing.

Progress Test 3 Listening scripts

Listening

Vaughn: Hi, Shelley.

Shelley: Hi, Vaughn. Just working on my presentation on the decline of the British bird population.

Vaughn: Birds, hmm.

Shelley: Did you know the number of countryside birds has decreased dramatically over the last two decades.

Vaughn: Er, can't say I'd noticed. I don't get out to the country much.

Shelley: Well, maybe you should, while there's still some birds there.

Vaughn: Well, Shelley, these things happen. We can't ...

Shelley: This is important, Vaughn. The status of the wild bird population is indicative of the health of the environment. A struggling bird population means a sick environment.

Vaughn: So, why is this happening?

Shelley: Mainly loss of habitat. The countryside is quite different from what it was twenty or thirty years ago.

Vaughn: Why's that?

Shelley: It's mostly due to changes in farming practices.

Vaughn: Really? The farmers are to blame?

Shelley: To an extent, yes. Certainly the rise in the use of chemicals has had a big impact.

Vaughn: And they're poisonous to birds, are they?

Shelley: Not directly, but their use results in the loss of plant and invertebrate food sources. Kill off what a bird eats, kill off the bird.

Vaughn: That's why it's called 'the food chain'.

Shelley: Exactly. And another key factor is that crop sowing times have changed. In the past farmers would harvest a crop of, say, wheat, in high summer, and leave the remains of the plant, the stubble, in the ground until the next spring, when they would plant new crops.

Vaughn: So what happens now?

Shelley: Now they remove the stubble and plant the new crops in the autumn. Because it's in the ground longer, the crop has more time to grow and they get a better yield.

Vaughn: Well, how does that affect birds?

Shelley: The stubble used to provide an ideal habitat for birds to forage for food during the harsh winter months.

Vaughn: Oh, I see.

Shelley: Other habitats are destroyed in other ways. For example, grasslands are drained or ploughed up. Hedgerows are removed to create bigger, more cost-effective fields, but the hedgerows provided a place to nest, protection from predators, and a good source of food, like berries, seeds, insects and so on.

Vaughn: So, are all birds equally affected by this? Are they all in decline?

Shelley: The RSPB has ...

Vaughn: Sorry, the what?

Shelley: The Royal Society for the Protection of Birds has drawn up three lists – red, amber and green – which categorise over 240 birds in terms of conservation concern.

Vaughn: Red, amber, green – like traffic lights!

Shelley: That's right, the red list contains forty species including those that are globally threatened or those whose population has declined rapidly in recent years.

Vaughn: What's on that list? Blackbirds?

Shelley: No, though the starling, which is related to the blackbird, has recently joined the red list, as has the house sparrow. One of the birds people are most concerned about is the skylark, which has been greatly affected by habitat loss.

Vaughn: So what about the amber list?

Shelley: That includes 121 birds with 'unfavourable conservation status', or whose population has declined moderately in recent years. That includes birds of prey, such as the kestrel and osprey, woodland birds like the cuckoo, water birds like the long-tailed duck and farmland species such as the lapwing.

Vaughn: And are all other birds on the green list?

Shelley: Not necessarily. It doesn't include those species that are not native to the UK – foreign species that have been introduced deliberately, such as the Canada goose – or those that have escaped from domestic cages, for example parrots.

Vaughn: I don't suppose it includes summer visitors either, you know, like swifts.

Shelley: No, they are included in fact. They are in this last category.

Vaughn: So, is anybody doing anything about all this?

Shelley: Thankfully, yes. The RSPB is campaigning to stop this loss of biodiversity by putting pressure on the government to invest money to rescue and protect threatened habitats.

Vaughn: But I thought you said that farmers were to blame?

Shelley: Well, the RSPB is calling on the government to reform the Common Agricultural Policy so that agricultural subsidies only go to environmentally friendly farmers.

Vaughn: I guess that would encourage them to change the way they do things.

Shelley: That's right. And there's a further initiative which encourages farmers to be part of the solution, not part of the problem.

Vaughn: What's that then?

Shelley: They're being asked to turn off their seed drills occasionally for a count of two seconds, when sowing winter cereals. This will leave small areas of land, about four metres square, that can be used by skylarks for nesting and feeding.

Vaughn: And will farmers do this?

Shelley: As an incentive, they're being offered five pounds for each skylark plot. The RSPB is aiming for two plots in every hectare of land.

Vaughn: And will this work?

Shelley: Hopefully. The RSPB reckons that if these empty patches are created on just one fifth of Britain's arable farmland, the decline of the skylark could be halted and then reversed.

Vaughn: Well, that's very interesting. Let's hope it works.

Shelley: Yes, let's.

Speaking

Part 2

The last film that I saw was a French movie called *L'homme du train*, starring Johnny Hallyday. In English, I think it's called *The Man on the Train*. I saw the film with a couple of friends of mine at the French Cultural Institute here. It was an unusual film about two men in the sixties who meet by accident in a small town in France. Johnny Hallyday is a stranger who arrives in this town by train late in the evening and is looking for a chemist that is open to get some medicine for his splitting headache. Here he meets a retired school teacher who has lived all his life in this town.

The film is very slow-paced which allows the audience to get to know the characters and see how the relationship develops between them. On the surface they are very different people – Johnny Hallyday wears a black leather jacket and turns out to be a bank robber, whilst the school teacher looks and acts like a traditional school teacher in a small country town. However as the film goes on, we can start to see that each of these two characters wants to lead the other person's life. The bank robber has a love of books and poetry whilst the teacher wants the excitement of living outside the law.
It's a great film which tells us a lot about people and how they are and how they want to be – it's also got some very nice comic touches. I'd certainly recommend it to anyone – just don't expect an action movie!

Part 3

Examiner: Do you think it's best to see foreign films in their original language with subtitles, or dubbed into the local language?

Candidate: I think it's so much better to see the movie in the original language as it communicates a real sense of where the characters are. It's so important to create an authentic atmosphere. Dubbed films just don't sound right.

Examiner: Many people think that most film adaptations of books are very limited. Would you agree?

Candidate: I would disagree. I've read *Lord of the Rings* and seen all the movies and couldn't say that one is better than the other. They do different things – a book allows you to use your own imagination to picture the characters and places. The films, however, can bring it all to life, giving us the film director's interpretation. If the director and actors are good – the film will be good.

Examiner: Do you think that English language films should have limited distribution in other countries to help promote a local, national film industry in each country?

Candidate: Um ... No, I don't think so. If people want to watch Hollywood movies, they should be allowed to watch them. There will always be a place for a local film industry because in the end people want to see films which also relate to their own culture. There is space for both.

Examiner: Many companies place their products in films as a way of indirectly advertising them. Do you think this practice should be banned?

Candidate: Absolutely! I hate it when you pay money to go to the cinema to watch a film and then find it is one long advertisement for a particular product. Why should I pay to watch an advert? The dangerous thing about this type of indirect advertising is that some people don't realise what is happening.

Practice Test

Listening

Section 1

Cosmic: Cosmic Home Delivery. My name is Gary. How may I help you today?

Customer: Hello. I'd like to place an order.

Cosmic: Certainly, madam. I'm afraid our computer system crashed earlier today. I'll have to take the details down on paper and then enter them later, when it's been fixed. Is that OK?

Customer: Yes, of course.

Cosmic: So, can I take your name, please?

Customer: Yes, it's Alexandra Hornby.

Cosmic: Sorry, could you spell the surname for me? H ...

Customer: Yes, then O-R-N-B-Y.

Cosmic: Oh, fine. And then your address.

Customer: That's number 28, Wood Road, which is in Ilford – I-L-F-O-R-D and that's near Northchester. The postcode's NC1, er, 2, er, FR.

Cosmic: Thank you. And do you have an account with us?

Customer: I do. I've got the number here – 9-double 4-5-6-7-8-1. Is that long enough?

Cosmic: Um, 8 digits, yes, it is. Good, now what would you like to order today?

Customer: I want a coffee table. I think there's only the one type.

Cosmic: I expect so. Perhaps you can tell me the price – I can use that to check later, just in case there's more than one.

Customer: Yes, it's 39 pounds 99p.

Cosmic: Fine. Sorry about all these extra questions.

Customer: It's no problem at all.

Cosmic: Now, that size of order value does mean you're entitled to a free gift. Did you want to take up that option on this occasion?

Customer: Yes, I do. I've already got a calculator like the one on offer, but I do like the look of the handbag, so I'd like one of those, please.

Cosmic: Certainly. And can I just check – as an account holder, you may have been sent a voucher ...

Customer: Oh yes, for a discount. Let me see ...

Cosmic: The reference number will probably start with the letter D.

Customer: It does, and it continues B-R-29.

Cosmic: Great. Now, how would you like your order delivered? There's Standard Service, within a week, and then Express, which comes within 2 days, or Special, which means it arrives the same day.

Customer: Hm, Special would be ideal, but I know it's a bit expensive, so I'll make do with Express, I think. Standard is very slow.

Cosmic: It is, to be honest. Well, that's your order completed. Could I just trouble you for another minute or so, to ask you a couple of questions to help us improve our service as much as possible?

Customer: Yes.

Cosmic: Firstly, we do like to try and keep a record of how customers have heard of Cosmic Mail Order. How did you? Was it from our advertising campaign?

Customer: Oh, I think I would have remembered any advertisements, on TV for example. I only read newspapers occasionally, so that would have passed me by. A friend of mine had been using you for years, and encouraged me to give it a go, so I looked on the Internet to find you, to see what was available on the site.

Cosmic: That's great, thank you. The second thing is, we're thinking of introducing a number of promotional offers ...

Customer: Oh yes, I got a little brochure about them with my last order. I've got the list here.

Cosmic: So can I ask which ones appeal to you?

Customer: Well, let's see. There's quite a few things here that don't really apply to me, actually. For example, my mobile phone bills are so low that I hardly notice them. Likewise, they shut down the local cinema. I do enjoy a meal out, though, so that discount could be of interest, and I like to get away at weekends when I can, and some of the places you want to visit are expensive, so it'd be lovely if they became cheaper. As for planes, well I haven't been abroad for a while now, and in any case, I prefer the train where possible.

Cosmic: Fine. Well, thank you very much for your time.

Customer: Not at all.

Section 2

Hello, everybody. It's nice to see that so many of you made it, even on an evening as rainy as tonight! OK, now whether you're new to cycling as a sport or are returning to it after some years' absence, I'll just go over a few basic points. Now, race preparation is a complex business, and there are many factors to consider.

The first area of concern is mechanical. Now this involves the machine itself, that is, your bike, and also, no less importantly in fact, clothing – this will protect your body and aid your performance, providing you wear the right kit. The next area to concern yourselves with is the mental. This is you as a person. You can have the best bike in the world, but you won't get the most out of it if you don't have the right mindset. So, tactics are important to consider. Another factor which is essential to a good performance is determination – you need to feel this so that you can really push yourselves to your limits. To accompany this, you also need knowledge of your bike, yourself, physics, other riders, the course, and so on. Finally, strange though it may seem after all I've said so far, you need relaxation. If you can't switch off sometimes, you won't get to re-charge your batteries.

OK, and that takes us on to the physical side of race preparation. The first and perhaps most obvious aspect of this is training, and we'll come to some of the details of that in a minute. Another important aspect to pay attention to is diet, and you'll soon find that if you don't eat well, you won't see yourselves performing as well as you might. Then there's also the question of style, and you'll need to learn to develop the most effective ways for each of you to deliver your performance – and the details of this depend on which kind of event you're competing in. And finally, you need to take body care into account. You need to stay healthy in order to be able to give of your best. Right, those are the basic ingredients of race preparation.

Now I'd like to turn your attention to some of the details of a good training regime, and what you should and shouldn't do through the year. Let's consider various activities in turn. The first thing to think about is circuit training. This is an indoor series of gym exercises, designed to work on all parts of your body. This starts as the racing season closes with the end of summer and continues right through the cold season, and stops you going off the boil when you aren't competing. It's extremely beneficial, although I would stress you do need a qualified gym instructor to tell you how to do it properly. Next, weight training. This is also very good for cyclists and it tops up the natural strength that cycling produces. Exercises need to be arranged as part of a carefully calculated routine, and this routine needs to be sensibly followed. It's a good idea, for continuity, to carry on using weights throughout the year, as you can lose strength just as quickly as you can gain it. Something simpler, and requiring no equipment, is mobility work. While cycling has a great many benefits for the body, it doesn't work every part of it, and indeed keeps some parts locked in pretty much the same position. So exercises that involve twisting and turning and generally promoting flexibility are advisable before the start of each race throughout the summer season. Another important activity is pleasure riding. Perhaps this doesn't sound so important to you, but you don't race all year, and when you do race, it's pretty hard riding – so you can sometimes forget that cycling is actually basically a fun thing to do. So, when you hang up your racing bike at the end of the summer, get out another bike and go for some gentler, enjoyable rides during the winter – weather permitting, of course. And finally, running. This is of course another sport in its own right, and for this reason some cyclists are rather sniffy about it. However, it is good exercise and maintains aerobic fitness very effectively. But it is rather hard on the knees, and in different ways from cycling, so you're best advised to keep your running to out of the racing season, and wrap up well against the cold when you do go. So, now I'd like to say a little bit about …

Section 3

Anna: So, Jane, Mark, we need to press on with the assignment.

Jane: Yes, we do, Anna.

Mark: And we need to begin by going to the best sources of information.

Anna: Right, now, we're looking into how films get altered sometimes if they're not going to do very well.

Jane: We need to know quite detailed things. So ordinary magazines, you know, leisure interest ones, may lack the detail we're after. Specialist magazines, on the other hand, will probably be helpful.

Mark: I think that's true.

Anna: OK, so we'll root out some of those. And what else?

Mark: There ought to be stuff available online …

Jane: Hmm, but we'll have to be selective.

Mark: Sure – not general cinema websites, you mean.

Jane: I think it'd be useful to go onto studio websites. Then we'd get pretty specific information.

Anna: Even if it might be a bit biased. Yes, OK. Anything else?

Mark: I think we also need to think about the point of consumption, so perhaps reviews would be good to look through.

Jane: OK, though we'll get most by looking at local reviews, I think.

Anna: I agree.

Mark: Good.

Anna: So far, so good. Now, we'll gather all that together – we can take one source each – and then …

Jane: I don't know.

Anna: About what, Jane?

Jane: I mean, OK, so we get all this stuff, probably loads of it. But do we know how to assess the usefulness of it? I know it'll all be factual, and so true in that sense, but we need to know what it can really tell us.

Anna: Perhaps we should cross that bridge when we come to it. Mark?

Mark: Yes, though I think Jane's got a point. But the problem for me is the context we're working in. We've got the assignment instructions, and in a sense it's all very straightforward – we know which parts are worth how many marks and so forth. But the focus seems to be heavily on the objective and quantitative side of things, whereas I would be hoping to be asked for more interpretative work. And there would have to be time to do that, within the framework we've been given.

Anna: Well, again I think we should just see how we go with it.

Jane: Shall we check that we're clear on just how films are altered?

Mark: Good idea. We'll list the various 'cures' that film-makers use when a film looks like it might be going to fail.

Anna: The first one is called 'tweak every joke'. Sometimes they change lots of the jokes after showing the first version to a test audience. Comedy is the only film genre with a reliable formula: the more the audience laugh, the more income the film gets.

Mark: And it's not very funny if your film makes a loss.

Jane: And the next one they call 'change the ending'. Occasionally they completely change the way the film ends. It seems a bit strange, maybe, but the problem is it's far from easy to know in advance what reaction you'll get from an audience. So if it turns out after all that the test audience doesn't like your ending, you've no choice but to do another one.

Mark: What's the third one?

Anna: That'd be 'fix the tone'. If, for example, your film begins as a kind of gentle comedy, but then gets too serious, or horror-like, audiences get confused.

Mark: So one thing that really matters is being consistent, you mean?

Anna: Yes, that's the logic, I think audiences don't want to have to switch track.

Jane: It's more complicated than people imagine, isn't it? Now, the next cure is a big one: 're-shoot'. This may seem drastic, but in some cases it's the only option available. It's very expensive, of course, but the film's backers will see this as protecting their investment.

Mark: Are there any others?

Anna: One more and it's another big one, or at least has the potential to be a big one. It's 'shift the genre', and this is done when the test audience seem to only like one half of the equation, as it were. Maybe you made a musical, but they only like your comedy story-line, not the songs, so you do it all again without the songs. It could broaden the film's appeal, get a wider audience.

Jane: It's weird to think how much extra work has gone into some films, isn't it?

Section 4

Now, in looking at the history of anywhere, we need to accompany our discussion of the facts with some consideration of what facts mean, or which facts have meaning. But more of that later. Let's start with looking at one very important period of the history of the Mediterranean – the period of Roman influence. This was born out of the death of Alexander the Great. His grip on the area went, and what directly ensued was a period of intense conflict which focused on the eastern Mediterranean, as opponents sought to gain control of that area. Rome began to emerge from this as a potential dominant force, but obstacles such as a lack of overall planning and in particular the ever-present menace of piracy lay in the way of success. Gradually, the Romans improved the power of their ships and fighting equipment. They also formed a series of alliances, which effectively reduced the size and number of enemies, and a key measure they took was creating an office of government specifically charged with repairing their fleets.

However, many ships they and their enemies had, Rome from now on would have the highest proportion out patrolling and fighting. There were of course no engines in ships in those days, and another aspect in the battle for supremacy was speed through human effort. Rivals built ships with ever-increasing numbers of oars but what really mattered was the amount of human pulling strength attached to each one, and the Romans, benefiting from their great population of slaves, was able to have every oar pulling faster and harder than anyone else's. Eventually, Rome's dominance was more or less complete. Their possession of the sea became their enjoyment of the sea. Their name for the Mediterranean translates as 'our sea' and that's how they saw it. The shores were of course by now very well fortified, but now the Romans began also to demonstrate their comfort in power, and put up a great many large houses, visible far out to sea. Wealthy merchants and retired generals lived in leisure in these temples to their own prosperity. The maritime security of the Mediterranean and the immense availability of trading destinations meant that all sorts of suppliers and craft producers clustered round any reachable section of the sea's edge. Some of these settlements were small, while others were large, producing salt fish in Italy, for example, or the fish sauce that was sent to every corner of the Empire from Spain. With dominance of the seaways established, the primary purpose of ships evolved from fighting other ships to transporting goods. Economics generated a drive to carry as much as possible on each trip, and as ships were built for carrying loads such as wine or stone exclusively, these types of specialised vessels led to greater cost-effectiveness. And so things continued for many years.

However, I referred before to the question of needing to consider how to approach history. Assumptions about what history is are as varied as the historians who have made the Mediterranean the subject of their studies. Three key historians have moulded, in different ways, our approaches to understanding the Mediterranean's past: Michel Balard, Fernand Braudel and Nicholas Horden. The first of these, Balard, felt strongly that an approach based on examining countries around the Mediterranean, as if they shaped its history, was inadequate. More was to be learned, he argued, by treating the Mediterranean as one of a number of seas, such as the Black Sea and the Atlantic Ocean, and thus integrated into trading routes that stretched from Madeira to Krakow. Our second historian, Fernand Braudel, took a rather different, though related, tack. He argued that the ways societies operate, right down to their individual members' actions, is subject to permanent conditions, such as whether they inhabit mountain, plain or sea. Features of the coastline and adjacency to the waters of the Mediterranean are thus the focus of his approach. And then we can add a third way of looking at history, or at least the history of the Mediterranean, and that's the one propounded by Nicholas Horden. His arguments are asserted quite strongly, not least because he sees a greater consistency across time than many other historians have allowed for. There are those who take ecological events, particularly catastrophes such as the volcanic destruction of Pompeii, to be the shapers of different sections of the Mediterranean's history. Horden says these are simply incidents, and that the major trends ride on through them. For him, while there may be cultural mutations and fashions in types of consumer goods, what matters is the system of trading goods that satisfy primary needs, such as grain, oil, metals and timber. So, for us, out of this complex of views, our job is to seek a synthesis and form a sophisticated approach – not just to the Mediterranean, but to any time or region in history.

Speaking

Part 1

Examiner: What are you studying?

Candidate: I'm doing a part-time marketing course at the moment.

Examiner: When did you decide to study that subject?

Candidate: I first had the idea about a year ago when my job changed and I had to get involved with promoting and presenting the services my company offers.

Examiner: What do you find most interesting about your subject?

Candidate: I think the human and psychological dimension of marketing is the most interesting – you know, the reasons why people choose certain products or services and what brands mean to people. I'm particularly interested in the relationship between a want and a need, for example.

Examiner: Do you enjoy being a student?

Candidate: Yes, but combining study with a full-time job can be a bit of a challenge!

Examiner: Is there anything you don't like about your studies?

Candidate: The sheer volume of reading can be overwhelming. Sometimes it's difficult to extract the salient points.

Examiner: Now, let's move on to newspapers and magazines. Which newspapers and magazines do you read?

Candidate: I read a national newspaper about once a week and always at the weekend, and I read the local paper nearly every day. I don't really read magazines anymore. I'm just too busy!

Examiner: Which parts of a newspaper do you think are most useful for people to read?

Candidate: It depends. Of course, business people will look at the business section, but if you're living abroad, you'll probably turn first to any news about your part of the world. Actually, if we're honest, the one thing probably everyone reads is the TV page!

Examiner: How can reading newspapers and magazines help in learning English?

Candidate: You can get access to a wide range of vocabulary and topics and maybe some cultural information. It's probably important to distinguish between being helped to understand English and to actually speak English. Newspapers help you understand more.

Examiner: Do you think that the Internet will replace newspapers and magazines?

Candidate: Not entirely – at least I hope not!

Examiner: Now, let's move on to music. Do you play a musical instrument?

Candidate: Well, I learned the violin and the piano when I was a child, but I wouldn't actually say I can play them now!

Examiner: What's your favourite kind of music?

Candidate: I like classical and jazz best.

Examiner: What kinds of music are the most popular where you live?

Candidate: Well, that probably depends on people's ages. Just round the corner from my house there's a club that plays a lot of folk music, but if you mean in my city, then you can hear more or less every kind of music.

Examiner: Do you think it's better to listen to live music or recordings?

Candidate: Well, probably live music because there is such a special atmosphere if you go to a concert, either classical or pop. But on a practical level, you can play recorded music any time, for example while you are driving. Going to a concert is a special occasion and it can be expensive too!

Part 2

Examiner: I'd like you to describe how you would like to spend a free day.

Candidate: Well, first of all I wouldn't set the alarm clock. I would sleep until I woke up naturally. Then I would make a cup of tea and take it back to bed and read or listen to the radio until I was ready to get up. Then I'd like to have a delicious breakfast and go out for a long walk along the river and into town. I'd do some shopping – for clothes – not food! I'd like to spend the rest of the day at home on the sofa – or in the garden if the weather was nice, reading, watching TV and talking to my husband. A yoga class and a nice dinner would round the day off perfectly!

Examiner: Do you think you might spend a free day like this soon?

Candidate: Yes, why not?

Part 3

Examiner: We've been talking about how you would like to spend a free day, and I'd like to discuss with you one or two more general questions relating to this. Let's consider, first of all, the leisure time that people have. Do you think that people generally have enough leisure time?

Candidate: No, not really. I think despite what was predicted a few years ago, people don't have all that promised leisure time and in fact, people are working longer and longer hours. It's a pity because some people really do live to work instead of work to live.

Examiner: What kinds of leisure activities do you think might be popular in the future?

Candidate: It's hard to say, but I would imagine as free time becomes more precious, people will want to treat themselves to expensive pastimes and hobbies. What I mean is people may not have time to take a whole day to go fishing or walking, but will want to do a short burst of activity. And of course, TV and computer games will still be around as leisure activities, I'm sure.

Examiner: Do you think that education influences the ways people spend their leisure time?

Candidate: I suppose so. Habits like reading or going to the theatre or being in conversations about the state of the world can be formed when you're a student and could stay with you later on. I'm not really sure ...

Examiner: What effects on society do you think any changes in the retirement age might have?

Candidate: If people work till they are older – say seventy – I suppose they will have less time to spend with their grandchildren! But seriously, older people have a lot of wisdom to give us, but if they're busy or not well because they're overworked, then that is a loss to society. If on the other hand, they retire earlier, there may be economic costs, but our society would be richer in other ways.

The Speaking DVD

About the Speaking DVD

There is a speaking DVD available for both *IELTS Express Intermediate* and *IELTS Express Upper Intermediate*. Each DVD is approximately 38 minutes long and features a complete candidate interview conducted by a qualified IELTS examiner. This section contains photocopiable activities for your students to do while they watch the DVD, and a transcript.

The DVD consists of five parts:

1, an overview unit, in which the examiner explains in detail the criteria by which a candidate's performance is assessed;
2, 3 and **4,** a description of each of the three parts of the IELTS Speaking exam, giving tips on performance appropriate to each section, e.g.

- Part 1: the examiner talks about extending responses and using 'fillers' such as 'sort of', 'you know' and 'anyway.'
- Part 2: the examiner talks about writing notes before beginning the Long turn and how to clarify questions.
- Part 3: the examiner talks about demonstrating a wide vocabulary and using conditional sentences.

5, a complete model interview: an uninterrupted recording of a full three-part Speaking exam taken by a candidate at the appropriate level.

Throughout, points are illustrated using clips of 'model candidates' of different nationalities performing a simulated IELTS Speaking exam. Some are examples of good language usage, some are not so good, but these 'model candidates' have all been specially selected to reflect the grade of each *Coursebook* and so provide a realistic level for students to aim for.

Using the Speaking DVD in your class

The *Speaking DVD* reflects and builds on what students have learned about the Speaking exam in the Speaking sections of the *Coursebook*. Although you could choose to show the DVD in instalments after students have been introduced to the relevant sections in the *Coursebook*, the DVD can also be used as a revision aid a few days before students attempt the Practice test. This would really help students to consolidate what they have learned throughout the course, but it is also useful because the Practice test uses similar material to the examinations on the *Speaking DVD*.

Ideally, if you have time, it would be advantageous to show the DVD twice. Once before the students take the *Coursebook* Practice test and once after, providing a good opportunity for reflection on what candidates need to demonstrate in order to achieve the grade they want at IELTS.

The accompanying photocopiable activities for each section should be given to students before they watch the DVD, for them to fill in as they watch. Answers are provided in the answer key. If necessary, the DVD transcript can also be photocopied and given to your students.

IELTS Speaking Module Overview

(I = Interviewer; E = Examiner)

I: We are here today to talk about the Speaking Module of the IELTS Exam and which skills and techniques students need to develop to be successful.
We asked an experienced IELTS Examiner and teacher, Ranald Barnicot, of Barnet College in North London, first to examine eight candidates at different levels from different nationalities, then to analyse their performances in each part of the Speaking module, and finally to assemble a range of extracts to illustrate these skills and techniques.
Tell me about the IELTS Speaking Module.

E: Well, it consists of three parts, which place rather different demands on the candidates, and it's always a one-to-one interview. However, there are general criteria which apply to all three parts.
There's fluency and coherence, to start with. This involves the ability to speak without too much hesitation or pausing to correct one's own mistakes. I mean, it's good to do that, but too much gets in the way of communication. And you should also speak at a proper speed so that you don't put a strain on the examiner by speaking too slowly or too fast.

I: Could you explain coherence a little more?

E: Basically, it means connecting your ideas in a structured way so that the listener can understand what you're trying to say. And this way you can also avoid repeating yourself.

I: And what about grammar? I imagine that's pretty important.

E: Yes, very much. Candidates' structures should be reasonably correct and they should show that they can use a lot of different grammatical structures, some fairly complex and not just the basic ones. We call these two factors accuracy and range. And, of course, the same things apply to vocabulary.

I: How about pronunciation? I imagine it's not very satisfactory having brilliant vocabulary if nobody can understand what you're saying.

E: Absolutely not! And not only do candidates have to get individual sounds right, like distinguishing 't' and 'th', /t/ and /ð/ for example, they also have to stress the right word in a sentence and use an appropriate intonation. Examiners want to see evidence that candidates can use appropriate intonation, for example whether they're asking a question or making a statement, and also expressing their feelings.

I: So if the intonation is rather flat, it can express lack of interest.

IELTS Speaking Module Part 1

E: In Part 1, the candidate has to answer a number of questions about themselves, their hopes for their future, their education or their jobs, any hobbies they may have or sometimes customs and lifestyles in their countries.

I: What should candidates focus on when answering these questions?

E: A major element is extended responses – that is, not just giving one-word answers or answering with just a simple phrase or sentence.
Let's show Danilo from Italy as a good example of this.

E: Hello. My name is Ranald Barnicot. Can you tell me your full name please?

Danilo: My name is Danilo Guerini.

E: Uh-huh. And what should I call you?

Danilo: Oh, Danilo, Danni.

E: OK, Danni. Where do you come from?

Danilo: I come from Italy, from the south of Italy.

E: Thank you. Alright, so, what I'm going to do first is ask you some questions about yourself. And I'm going to start with your studies. What are you studying?

Danilo: At the moment I'm doing a foundation year at Barnet College because I would like to go to university hopefully next year. And I am studying IELTS, an IT course, and a general English course.

E: What do you hope to study in university?

Danilo: I'd like to study European studies, which involves politics and international relations, and, yeah, politics and international relations.

I: Yes, I see what you mean. He didn't give much information about himself or his part of Italy to start with but he does go on to describe his studies and what he wants to do in some detail. And I'm quite impressed by his vocabulary.

E: Yes, 'involves', for example, is quite a formal word but it's very useful in an academic context, or talking about one's studies, as Danilo is. I have to say, however, that he becomes a bit less fluent at the end.

I: Yes, he hesitates a little too much, doesn't he?

E: Daniel is another good example. Notice also the way he becomes more involved in the conversation by smiling.

E: Let's talk about music. Do you play a musical instrument?

Daniel: No, never.

E: What is your favourite kind of music?

Daniel: It could be folk or country – I love that, so in my country it's very popular, the folk, folk music.

I: Yes, but I'm reminded about what you said about intonation. It is a bit flat in Daniel's case, isn't it?

E: Candidates should have a rising and falling intonation. It makes them sound interested and engaged in the conversation. Very flat intonation can make the candidate seem bored and uninterested and can have a negative effect on the listener.

I: Is there anything else you can suggest?

E: Fillers. It is important that candidates use these to fill the gaps in conversation. They also buy them time to think. Phrases like 'anyway' or 'you know', 'actually', etc. Here are some common ones. First of all, Boya:

Boya: 'Cos I wanted to 'cos I'd like to be a producer in the future and to do some programmes for people.

E: What do you find most interesting about your subject?

Boya: I think is – produce programmes – the process when, when I do it, and I think I can find some interesting things in it.

E: Boya uses 'erm ... I think ...' Now watch these clips of Panille.

E: What are you studying?

Panille: Well, right now I'm studying English. It's an English foundation course. It's a one-year course. Um, yeah?

E: Why did you decide to study that subject?

Panille: Well, I want to become an actress and I want to study here in London, so I want to perfect my English, and I thought this would be a good course, and also I get practice in taking IELTS which will show what level my English is at.

E: What kinds of music are most popular where you live, in Norway?

Panille: Um ... Probably American and English pop music, yeah.

I: Well, in the first one, she uses 'well' a lot. And in the second she uses 'yeah'.

E: The 'yeah' is interesting – she's using it to show that she's finished speaking.

IELTS Speaking Module: Part 2

E: In this part the examiner asks the candidate to talk for one or two minutes about a topic, usually about their personal experience, something they've seen or done, somewhere they've lived, a person they know or a thing they own or would like to own – this type of thing.

I: Do candidates choose their own topic?

E: No, the examiner gives the candidate a card about a specific topic with four points that he or she must cover. The candidate then has one minute to prepare what they want to say. In fact, many students use this time to make notes; it helps them organise their ideas.

E: OK, let's move on to the next part. I'm going to give you a topic to talk about for one to two minutes. Before you start you will have one minute to write some notes if you wish. And, first of all I'm going to give you some paper and a pencil for your notes. And, here is your topic. I'd like you to describe how you would like to spend a free day.

E: And this is Daniel's card.

E: So the candidate has to explain how they would like to spend the day, what they might do, and who with. Also they must give a reason or reasons.

E: Of course, the candidate may need clarification about what he or she has to do, and Daniel asks for it:

Daniel: OK. Can I ask you, can I choose everybody who I want to spend the day with?

E: Yes.

Daniel: Alright. Thank you.

E: As does Panille:

E: I'd like you to describe how you would like to spend a free day.

Panille: OK. Is it like here in England, at this time, kind of?

E: Any time you like.

Panille: OK.

I: So Daniel asks if he can start by describing the person he wants to spend the day with and Panille wants to know if she has to talk about a specific time or place.

E: That's right. If a candidate is not sure about the instructions, it's a good idea to ask – that way they can be more confident that they are doing the task correctly.

I: What kind of notes should the candidates write?

E: Well, there are two main types of notes. Firstly, there's linear – you just make a list of points to cover – and then there are word webs as they're sometimes known. Candidates can of course write whichever kind they want, but many people prefer the second kind because they say it's easier to think of ideas that way. You start at the centre and move outwards. One thing which is helpful is to include a list of keywords, vocabulary which is important and which will impress the examiner.

I: Any advice for when candidates start speaking?

E: Yes, sound confident and look at the examiner. It helps sometimes to give a brief summary of what you are going to do.

Boya: Right – I'm going to tell you about spending a day with my parents.

E: It also helps, when you are writing your notes, to visualise the scene. What do you think of Panille's description?

Panille: OK – well I would like to go to – get up early, and go to a park like Hyde Park or something and have a picnic with my friends. And we would stay there all day, and everyone would bring something to the picnic and we would just sit there and talk and have fun – and I have some friends that play the guitar, so they would play the guitar, and we'll just talk and stay there until late and have a barbecue as well.

I: Yes, we can certainly imagine the sort of day she is describing. It's well visualised.

E: And what about her vocabulary?

I: A bit simple?

E: Yes. It's good that she knows specific words like 'barbecue' and 'picnic', but otherwise the words she uses are very high-frequency items, and the repetition is noticeable: 'talk' and 'stay there' are each used twice. What about Danilo?

Danilo: We usually have lunch out when, when I have free time and when I have a free day and after having lunch, we carry on shopping around and having a stroll.

I: He uses 'stroll' and 'shop around'.

E: Yes, so all in all he's got a good range of vocabulary, from the formal word 'involves' which we noticed earlier to these rather informal expressions.

E: Let's hear their comments afterwards about this part of the exam:

Panille: OK, so, what I found helped me was that I visualised the scene when I was talking about my day, my free day and I could actually see myself sitting in the park, having a picnic and then I could kind of grasp the details of what I wanted there, so that really helped me.
And, what I didn't do was write down keywords and vocabulary and I found that I repeated myself quite a lot, because I didn't do that and also I kept my language quite simple. I didn't say any big words, so, and I thought I should have.

Danilo: And ... also writing keywords helped me remember what I had to say like 'having a stroll' – writing, writing it down helped me to remember it, so I think it's really helpful.

I: I'd quite like to hear a whole Part 2.

E: All right. Here's Danilo.

Danilo: Alright. When I have a free day, I like to go to central London. I usually go with my girlfriend. We usually go to Camden Town where there's a big market, actually there are two or three markets. I like shopping around and looking at the different things there, they show there. I also like going to Oxford Street and Leicester Square – they are really different from Camden Town. And ... That's what I like of London, it's really different. You can find different, like two or three cities together.
We usually have lunch out when, when I have free time and when I have a free day and after having lunch, we carry on shopping around and having a stroll. I like to go to the centre because I just want to have a relaxing time and not thinking about working or studying or doing anything else, just doing what I want to do, really.

E: OK, so, do other people that you know like to spend the day in the same way?

Danilo: Yeah, I've got some friends who I work with – but we usually go out during the night, the evening. We go to, to some pubs in central London. Sometimes we go shopping or just looking around.

E: Well, that seemed to be pretty well organised. What did you think of his grammar?

I: A good range and quite accurate.

E: Yes, he uses 'after having lunch', 'some friends who I work with', so he uses quite complex forms. What about his pronunciation?

I: Very clear but, again, rather monotonous intonation. Perhaps he needs to sound more lively!

E: Yes, as we mentioned before, speaking with a rising and falling intonation pattern will make you seem interested and engaged in the conversation.

IELTS Speaking Module: Part 3

E: Part 3 is a question and answer session, but more of a discussion, based on the topic of Part 2. Candidates have to be able to cope with questions about the environment, educational system and, in general, issues facing either their country or the world.

I: So it's more abstract.

E: Yes, and it helps to know what is going on in the world.

I: What if a candidate doesn't understand a question?

E: Well, just like in Part 2, he or she should request clarification. Let's see how Boya does this.

Boya: Now people want to have more higher educations, 'cos they want to get good jobs, so they don't have much more time for leisure.

E: What effects on society do you think any changes in the ages at which people retire might have?

Boya: Sorry?

E: Well, there's talk about changing the law about retirement age, making it possible for people to retire later.

Boya: OK.

E: What effects will this have on society?

Boya: Oh – so you mean is the retired age for people in the society?

E: Yeah.

Boya: What I know in China is people probably at age 50, they probably will retired.

I: What will candidates be asked to do?

E: To express an opinion, appropriately introduced by 'I think' or 'In my opinion' etc and backed up by evidence. Let's see what Danilo uses.

E: Do you think that people generally have enough leisure time?

Danilo: No, I prefer to think that nowadays people don't have a lot of free time, and they ...

E: He says 'I prefer to think'. Now Boya:

E: Talking about the opposite of leisure, that is work, do you think that employers should encourage employees to work overtime, or discourage them from doing so?

Boya: I suppose not to do that 'cos it's better to encourage them, to make them, let them know what they ought to do, and they can do it by themself to work harder and make them work better, yeah.

E: She uses 'I suppose', but we are looking for a wide range of expressions such as 'I really believe' or 'What I think is'. Another thing a candidate may have to do is evaluate, to make a comment or decision on how valuable something is, how important or useful.

E: So do you think that the change has been for the better or for the worse?

Daniel: I think for worse because of relationships and ... Anyway a few people had contact with nature, and they just spend the time with new technology, like I said, computers, and television.

I: So Daniel doesn't think much of the way we enjoy ourselves these days. Evaluation, I suppose, is another way of expressing an opinion. Is there any other way that they will have to use language?

E: They might have to compare and contrast. Danilo's response is rather good on this:

E: How do you think the ways in which people spend their leisure time have changed since a hundred years ago?

Danilo: I think it's, it's changed a lot. Maybe 100 years ago people did not have, did not use to have weekends off or any week days off so they did not have much free time to spend in doing some leisure activities, whereas now life is much easier. People have two days, or at least one days off, so they can spend a lot of time doing relaxing activities, such as going to cinema, having a walk from time to time, or go shopping around, shopping around when they have a chance.

E: He uses 'whereas' to express contrast. Daniel does answer this question but his response is very disjointed and it's not easy to see the grammatical structure.

Daniel: What the difference is depends because of technology was developing, and so maybe a hundred years ago without television, and they without DVD, without CD, without computers, they spent a lot of time with, with their family and now ...

E: Another thing a candidate may be called on to do is to hypothesise.

I: Hypothesise?

E: Yes, to imagine a situation which doesn't exist in reality, to imagine what would happen if things were different from what they are. Let's watch an example.

E: Panille, how do you think people would react if there was no TV?

Panille: Well I think they would be quite shocked – because well, I never had a life without TV so I would be quite surprised and it would be shocking, shocking news. What would be a problem would be that we're used to getting news and stuff really quickly and we wouldn't be able to do that. I mean you wouldn't get to know that the TV was out before the newspaper came out today after and that would be a bit late, because today society is moving quite quickly, so that would not be good. Also, if it happened like for a long period of time you would probably go back, kind of back in time and start playing games again with the families and go out and see the blue sky and take some fresh air, and we would probably go back to what we used to do, going to the theatre and playing football with friends even though we are grown up and don't usually do that any more. But I think people would panic a bit, and also a lot of people would lose their work so, yeah, so many people would be out on the dole and that's not good.

I: Yes, she used 'would' consistently, didn't she?

E: In a second conditional construction. Now listen to Daniel:

E: What do you think would happen if the government banned overtime?

Daniel: Oh, it's terrible – if they had banned overtime, I think that people would do black business, we call that black business, so I think that employers would used the employee for overtime so I think that it's, it's useless to ban that.

E: Now let's listen to Boya.

E: What would happen if the government said people can work as long as they wish?

Boya: No, I don't think it's a good, good thing, 'cos people have rights and they don't want to work all their lifetime. They have time to relax, to leisure, to spend with their children or with their parents. So I think it have a retired age to, for people to retire, yeah.

E: She doesn't really answer the question, does she? She thinks I mean people should be forced to work all their lives, whereas I suggested that they should be allowed to.

E: Then there's something else they might have to do: speculate about the future. What form does Danilo use to do this?

Danilo: I think it might be a problem if you start going to a gym or anyway taking a gym sports – just going to a gym or swimming or running.

E: OK, so do you think that more and more people will take this up?

Danilo: Yeah, it's already, I think it's already happening now. Many people go to a gym before going to work, maybe in the, in the first morning, about seven o'clock, or after they worked.

I: He uses 'might'.

E: Yes, though that's also the form I use in my question. He might also have used terms like 'maybe', 'perhaps', 'in all probability', 'what seems likely is', etc.

I: And any other language they should concentrate on?

E: Yes, they need to use 'signpost words', words which indicate what is ahead or coming, what kind of thing the speaker wishes to say, perhaps reason or result, or an example, or to point out some kind of contrast.

E: Let's listen to Panille.

E: So, do you think the quality of life these days is worse than it used to be?

Panille: Well, the quality of life is definitely better in, in theory because now we live longer, and we don't, we are not as sick as they used to be. If we get a disease now you can be cured.

E: She uses 'because'. However, this is a very common word, so we're looking for terms like, 'since', 'as', 'seeing that' or 'the reason is'. For result, of course, there's 'so', or 'so that' or verbs like 'cause' or 'lead to'. And in this clip, Panille uses some quite advanced vocabulary.

E: Of course entertainment is a very big industry these days. What kind of impact does that have on our lives?

Panille: Well, I think all people are in some way or another are affected by their entertainment industry ...

I: She uses 'affected by'.

E: Yes, and she could have also used words like 'impact' or 'influence'.

E: Now in this clip Daniel introduces an example:

E: Let's take the opposite of leisure, that is work. Do you think employers should encourage employees to work overtime, or discourage them from doing it?

Daniel: I think that they could be, encourage them but it depends on the work as well, because, for example if I do my work for my, during my, my normal hours, so I am not able to do everything, and I will be, and I will do hard work, so I have to, I have to take overtime as well, so I will expect from my boss that he will encourage me or maybe he will help me.

I: Yes, he uses 'for example', but we can also use 'such as' or 'for instance'.

E: Or 'an example is ...', 'let's take as an example', etc.

I: You also mentioned 'contrast'. Danilo used 'whereas' earlier to express this.

E: Yes, what does he use here?

E: Do you think that people generally have enough leisure time?

Danilo: No, I prefer to think that nowadays people don't have a lot of free time and many people are workaholic and they prefer to spend their time working but in the other hand, there are people who like to have more leisure time but they can't because they, they have a lot of fees to pay, so they spend more time working rather than in free time.

I: 'On the other hand', except he said 'in the other hand'.

E: Exactly. Something else, although we haven't got any examples, is to outline your response. For example, 'There are two things I can say about this ...'.

IELTS Speaking Module: Sample interview

I: It might be helpful to watch a whole interview now.

E: Well, here's Boya. As you watch, think about those general criteria: fluency and coherence, vocabulary, grammatical accuracy and range, and pronunciation. To take the last point, how does she pronounce the word 'busy', B-U-S-Y, in Part 2? Does she make an effective use of stress and intonation? But, above all, is she clear enough? In Part 1, are her responses extended enough or too brief? Does she use fillers to give herself time to think?

I: Yes, and in Part 2, does she continue as confidently as she started?

E: Also, maintaining eye contact is very important, not staring at the examiner but glancing towards him on keywords which she wishes to stress.

I: And in Part 3 I'll look out for how she expresses her opinion, or compares and contrasts, speculates, or hypothesises.

E: And don't forget the use of signposting.

E: Good afternoon.

Boya: Good afternoon.

E: My name is Ranald Barnicot. Can you give me your full name please?

Boya: My full name is Boya Liu.

E: I see. And what can I call you?

Boya: You can call me Boya.

E: And where do you come from?

Boya: I come from China.

E: Well, I'm giong to start by asking you some questions about yourself.

Boya: OK.

E: First of all about your studies. What are you studying?

Boya: Now I am studying diploma in foundation studies in Metropolitan University.

E: Hmm-mm. And what do you intend to study at Metropolitan University in future?

Boya: Probably media.

E: Why did you choose that subject?

Boya: 'Cos I wanted to 'cos I'd like to be a producer in the future and to do some programmes for people.

E: OK. What do you find most interesting about your subject?

Boya: I think is – produce programmes – the process when, when I do it, and I think I can find some interesting things in it.

E: Do you enjoy being a student?

Boya: Yeah.

E: Is there anything that you don't like in your studies?

Boya: Oh, homework!

E: Let's move on to talk about newspapers and magazines.

Boya: OK.

E: Which newspapers and magazines do you read?

Boya: You mean in English or Chinese, or ...

E: In English.

Boya: OK. In English when I stay in London, probably read 'Metro' 'cos it's free and I can take it from the Tube.

E: Which part of a newspaper do you turn to first?

Boya: Probably entertainment.

E: Why?

Boya: 'Cos I don't like to be something very political things. I want to see something very relax. Yeah.

E: Is there any part of a newspaper that you don't read?

Boya: Things about politics.

E: I see. OK. Let's move on to music. Do you play a musical instrument?

Boya: Yeah.

E: What do you play?

Boya: Piano. Yes.

E: OK. What is your favourite kind of music?

Boya: All kinds if I like it, I think it's all beautiful music. But especially is classic, yeah.

E: What kinds of music are most popular where you come form?

Boya: Now in China is pop music, yes, and lots of pop stars singing pop songs.

E: OK. Do you think it's better to listen to live music or recordings?

Boya: I think for me, I think it's better to listen to record music 'cos it's better, but I think if you want to be a good singer it's better to have a live music, and it can show your, show your skills of the music and your voice.

E: Let's move on. So the next part I'm going to ask you talk about a topic for one or two minutes. Firstly you will have one minute to prepare the topic and write some notes if you wish. Here is some paper and a pencil for your notes. And this is your topic, 'Describe how you would like to spend a free day.' Alright?

Boya: Yeah.

E: Remember you have only one to two minutes so don't worry if I stop you. Would you like to start now, please?

Boya: Right. I'm going to tell you about spending a day with my parents. When I have a free day I love to be at home with my Mom and Dad, 'cos I think, sometimes when I stay at home it's very little time with them, 'cos I have to go to school and these free days when I'm with my Mom and Dad I can talk with them, chat with them and show what was the life in my schools, and let them know what happens and what I, now and what's different between now and the past and I think it's a good way to improve the relationships with parents like these. And after that I like to listen to the music and to relax myself, 'cos you know that studies in school are very busy and also very boring, so I want to do some things really can relax me and just forget things in the school. And why I would like to spend the day like these is first of all it can relax myself and secondly it can improve relationships with my parents and thirdly it's, I think it's good for health going out instead of staying home, so that's all, yeah.

E: OK. And do you think you will be able to do this soon? To spend the day with your parents as you said?

Boya: No, I don't think so 'cos I am in London and they are in China, yeah.

E: So, we've been talking about, we've been describing, how you would like to spend a free day. I'd like to ask you some more general questions on the same theme. These days do you think that people generally have enough leisure time?

Boya: I don't think so, 'cos these days are more competitive before, than before, so people more getting to spend time in work ... much more money to make the life better, yeah, so I don't think they have enough time for them to leisure.

E: OK. What kinds of leisure activities do you think might be popular in the future?

Boya: In the future, I think, probably stay at home, yes, 'cos they stay at home and do some things they want – don't do things somebody let you to do ... it's just, just relax, yeah.

E: Do you think that we need technology in order to use our leisure time effectively?

Boya: Technology ... I don't think so, 'cos, you know if you, if you use technology to leisure probably you will use your brain or use your body to do some sports. I think leisure is just relax, just stay at home and lie on the sofa and watch some TV or something like that.

E: OK. How do you think people used to entertain themselves, say, 100 years ago?

Boya: Hundred years ago – oh, is so long time ...

E: When they didn't have television or, or, or CD players.

Boya: Probably chat around with people they live with and stay at home and do something they like and if they are not, like to chat with people, just stay at home, yeah.

E: Do you think that education influences the ways that people spend their leisure time?

Boya: Education – for me I think, yes, 'cos you know now people want to have more higher educations, 'cos they want to get good jobs, so they don't have much more time for leisure.

E: OK. What effects on society do you think any changes in the ages at which people retire might have?

Boya: Sorry?

E: Well, there's talk about changing the law about retirement age, making it possible for people to retire later. What effects will this have on society?

Boya: Oh – so you mean is the retired age for people in the society?

E: Yeah.

Boya: What I know in China is people probably at age 50, they probably will retired. But I don't think is a goo[d] age, 'cos you know now that life better and better, so the life expense, expectancy is longer, so I think it can be 60, is better to retired.

E: What would happen if the government said that people can work as long as they wish?

Boya: No, I don't think it's a good, good thing, 'cos peopl[e] have rights and they don't want to work all their lifetime. They have time to relax, to leisure, to spen[d] with their children or with their parents. So I think it have a retired age to, for people to retire, yeah.

E: OK. Going back to entertainment. Are there any forms of entertainment that you disapprove of?

Boya: I disapprove of, is ... you mean is, I don't like it?

E: You don't like it.

Boya: I don't like playing video game.

E: Why not?

Boya: 'Cos I think is ... waste of time, and otherwise, but like watching TV.

E: Thank you very much. Thank you for speaking to us today.

Boya: Thank you.

E: Finally, you might like to hear Daniel, Danilo and Boya's comments on their performance.

Daniel: It was not bad but I found the third part most difficult, maybe because I had to answer your question immediately. And the first one was much easier, because the answer, the questions was simple. So maybe the second part was, was fine, because I could decide and maybe one minute was enough for me to do some work, and some notes, so that was fine.

Danilo: Actually, I found more harder the last part, the questions about the topic, 'cos they were really detailed and you have to think about, about them in a really deep way – you have to answer in a completed way so I definitely found them harder than the first ones.

Boya: In Part 3 and I think some questions are very difficult to understand, while, but I think it's better to ... we ask ... well you mean, it's like that. Or just to relax with the examiner and then make sure the question what it's mean, yeah.

I: Thank you for this most useful explanation, Ranald.

Overview

1 Listen to the examiner speaking about fluency and coherence. Which of the following does he talk about? Mark them with a tick.

connecting ideas .....
avoiding repetition .....
body language .....
appropriate speed .....
not hesitating too much .....

2 Are the following statements about accuracy and range **true** or **false**?

a It's best to use a limited range of grammatical structures correctly. True/False
b All IELTS candidates should try to use complex vocabulary when possible. True/False

3 Listen to the examiner speaking about pronunciation, and complete the following paragraph.

> Not only do candidates have to get .... right, they also have to .... the right word in a sentence and use an appropriate intonation. For example, to show whether they are .... or .... and also for expressing their ....

is page may be photocopied.

Part 1

1 According to the examiner, in Part 1, candidates will answer questions about themselves – their hopes for the future, ...

....

....

or .... and .... in their countries.

Giving extended responses

Danilo

2 What does the examiner say is good about Danilo's answers?

....

....

....

3 What does the examiner say needs to be improved in Danilo's answers?

....

....

....

Daniel

4 What does the examiner say is good about Daniel's performance?

....

....

....

5 What does the examiner say needs to be improved in Daniel's answers?

....

....

....

Fillers

6 What three examples of fillers does the examiner provide? ....

....

....

7 What fillers do these speakers use?

a Boya ....

b Panille ....

Pairwork

Now speak with a partner to ask and answer the following questions. Remember to use fillers and to give extended answers.

Where do you come from?
What are you studying?
Why did you decide to study that subject?
What do you find most interesting about your subject?
Do you play a musical instrument?
What is your favourite kind of music?
What kinds of music are most popular in your country?

This page may be photocopie

Part 2

1 Which of the following are you unlikely to speak about in Part 2?
a a person you know
b politics in your country
c an experience you have had

2 Are the following statements about Part 2 **true** or **false**?
a Candidates are given a selection of cards to choose from. True/False
b Each card contains four points. True/False
c You have two minutes to prepare. True/False

Asking for clarification

3 What language does Daniel use to ask for clarification?
...., can I choose everybody who I want to spend the day with?

4 Panille asks if she has to talk about a specific .... or ....

Candidate notes

5 What two types of notes does the examiner mention?
.... and ....

6 Which type does he say makes it easier to think of ideas?
....

7 It is helpful to include a list of ....

Advice for starting speaking

8 What three pieces of advice does the examiner give?
a sound ....
b .... the examiner
c start with a .... of what you are going to do.

Panille

9 Watch Panille's Part 2 talk and listen to her talk about her performance. Answer these questions.
a What does the examiner like about Panille's Part 2 talk?
b What does he criticise?
c What did Panille think was helpful for her?
d What does she think she could have done better?

Danilo

10 Watch Danilo's entire Part 2 talk and answer these questions.
a What does the examiner like about Danilo's Part 2 talk?
b What does he criticise?

Pairwork

Now practise Part 2, using the Topic Cards below. Firstly, spend one minute preparing by writing notes. Then, speak for one or two minutes on the topic given. When your partner has finished, ask the bridging questions below.

Student A

Describe a beautiful or interesting natural feature (e.g. mountain, river, lake) that you have visited.
You should say:
what and where it was
who you went with
how you got there
and how you felt about it once you arrived.

Student B

Describe something (e.g. book or website) that has helped you to learn something (e.g. the English language).
You should say:
what it was
how often you used it
how it helped you
and how you felt about using it.

Bridging Questions

Student A

Ask Student A one or both of the following questions:
Do you think that you will go there again?
Is it popular with visitors to your country?

Student B

Ask Student B one or both of the following questions:
Do you know other people who have found it useful?
Do you think it will continue to be used in the future?

This page may be photocopied.

Part 3

1 Questions in Part 3 are more .... than the other parts of the exam.

Asking for clarification

2 What does Boya say to ask for clarification?

....

Expressing an opinion?

3 What phrases used to express an opinion does the examiner suggest?

Danilo

4 What phrase does Danilo use to express his opinion?

Boya

5 What phrase does Boya use to express her opinion?

6 What other phrases does the examiner say she could have used?

Evaluating

7 Evaluation asks you to decide how .... something is.

Comparing and contrasting

8 What phrase does Danilo use to express contrast?

9 Which speaker compares and contrasts more effectively? Danilo/Daniel

Hypothesising

10 What word does Panille use to show she is hypothesising?

....

Speculating about the future

11 What word does Danilo use to speculate about the future?

....

12 What other expressions could he have used?

....

....

Sign post words

13 What are the purposes of signpost words. Give two examples. ....

14 What synonyms are given for the following signpost words?

a because ....

b for example ....

15 What expression does Danilo use to show contrast?

....

Pairwork

Now ask each other the following questions from the DVD.

Student A

What effects on society do you think any changes in the ages at which people retire might have?
Do you think that employers should encourage employees to work overtime, or discourage them from doing it?
What do you think would happen if the government banned overtime?
What would happen if the government said people could work as long as they wished?
Do you think the quality of life these days is worse than it used to be?

Student B

Do you think that people generally have enough leisure time?
How do you think the ways in which people spend their leisure time have changed since a hundred years ago?
How do you think people would react if there was no TV?
Entertainment is a very big industry these days. What kind of impact does that have on our lives?

This page may be photocopied.

Sample Interview

Watch Boya and evaluate her performance by circling the appropriate number:
1 = good; 2 = OK; 3 = needs improvement. Provide examples where possible.

Exam section	Evaluation			Reason(s)/Example(s)
Part 1				
Use of fillers	1	2	3	
Extended responses	1	2	3	
Part 2				
Interaction with examiner	1	2	3	
Request for clarification	1	2	3	
Organisation of talk	1	2	3	
Ability to keep going	1	2	3	
Part 3				
Understanding of questions	1	2	3	
Request for clarification	1	2	3	
Evaluation	1	2	3	
Comparing and contrasting	1	2	3	
Hypothesising	1	2	3	
Speculation	1	2	3	
In general				
Fluency and coherence	1	2	3	
Vocabulary	1	2	3	
Grammatical accuracy and range	1	2	3	
Pronunciation	1	2	3	

Pairwork

Use the questions from Parts 1, 2 and 3 to practise the entire interview in pairs. Assess your partner's performance according to the criteria in the table above.

This page may be photocopied.

Overview

1 connecting ideas; avoiding repetition; appropriate speed; not hesitating too much
2 a False; **b** True
3 individual sounds; stress; asking a question; making a statement; feelings

Part 1

1 their education, their jobs, their hobbies
2 he gives extended responses, his vocabulary
3 he becomes less fluent toward the end, he hesitates too much
4 he becomes involved in the conversation by smiling
5 his intonation is flat
6 anyway, you know, actually
7 erm ...; I think
8 well; yeah

Part 2

1 b
2 a False (there is no choice); **b** True; **c** False (one minute)
3 'Can I ask you ...'
4 Time; place
5 Linear; word web
6 Word web
7 keywords
8 a confident; **b** look at; **c** brief summary
9 a her talk is well visualised; **b** her vocabulary is a bit simple, and there is noticeable repetition; **c** visualising the scene; **d** she should have written down keywords and vocabulary.
10 a it was well organised; good range and accuracy of grammar; clear pronunciation; **b** monotonous intonation

Part 3

1 abstract
2 Sorry?
3 'I think'; 'In my opinion'
4 I prefer to think
5 I suppose
6 I really believe; What I think is ...
7 valuable/important/useful
8 whereas
9 Danilo
10 would
11 might
12 maybe; perhaps; in all probability; what seems likely
13 indicate what is ahead or coming; firstly/secondly etc
14 a since; as; seeing that; **b** for instance; such as; an example is; let's take ... as an example
15 on the other hand

Sample Interview

Part 1

Use of fillers: 2 – erm ... /OK/yes
Extended responses: 2 – does elaborate at times, but responses are often minimal; she sometimes needs prompting

Part 2

Interaction with examiner: 1 – glances at the examiner and smiles
Request for clarification: does not apply
Organisation of talk: 1 – announces topic, moves smoothly through phases, uses some signposting: 'Firstly, ... Secondly, ...'
Ability to keep going: 1 – very little hesitation

Part 3

Understanding of questions: 2 – generally OK, but doesn't understand question about retirement age
Request for clarification: 3 – 'Sorry?'; 'I disapprove of, you mean is, I don't like it?'; (can't manage sentence structure)
Evaluation: 2 – responds adequately to question about technology ('I don't think so, because these days ...')
Comparing and contrasting: does not apply
Hypothesising: 3 – invited to hypothesise in response to question about retirement age, but doesn't misunderstand the examiner's question
Speculation: 3 – doesn't use future tense or any appropriate form

In general

Fluency and coherence: 2 – keeps going, but delivery disjointed at times through language difficulties
Vocabulary: 3 – rather narrow range
Grammatical accuracy and range: 3 – a fair range of grammatical forms but a lot of errors; need to link clauses more ('I want to do things really can relax me.')
Pronunciation: 3 – very clear, although rather staccato; stresses keywords quite effectively; pronounces /iː/ instead of /ɪ/ (e.g.: /biːzɪ/ for /bɪzi/)

Describe a time when you lost something.

You should say:
what you lost
where you lost it
what you did when you realised it was missing

and explain how you felt.

Describe an interesting journey you have taken.

You should say:
where you were going
how you travelled
who you were with

and explain why the journey was interesting.

Describe a teacher you respect.

You should say:
who this person is
when you met this person
what he or she taught you

and explain why you respect him/her.

Describe an enjoyable party you have been to.

You should say:
what the occasion was
where the party was
what happened at the party

and explain why it was so enjoyable.

This page may be photocopied.

Describe one of your favourite items of clothing.

You should say:
what the item is
how long you have had it
how often you wear it

and explain why you like it so much.

Describe a house you particularly like

You should say:
who lives there
where it is
what it looks like

and explain why you like this house so much.

Describe one of your favourite songs.

You should say:
what the song is
who performs it
what it is about

and explain why you like this song.

Describe one of your favourite dishes.

You should say:
what the dish is
how it is made
how often you eat this dish

and explain why you enjoy this food so much.

This page may be photocopied.

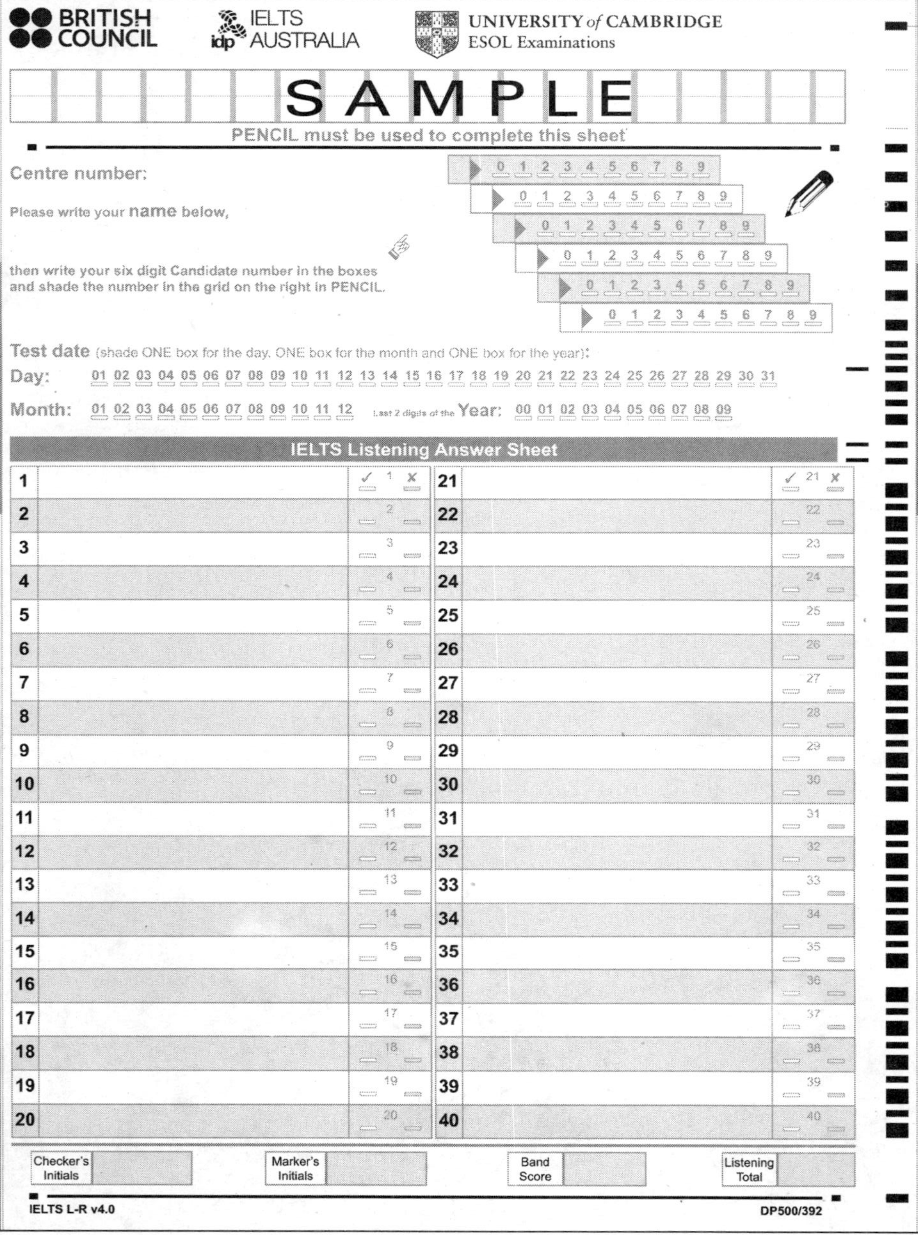

BRITISH COUNCIL | IELTS AUSTRALIA | UNIVERSITY of CAMBRIDGE ESOL Examinations

SAMPLE

PENCIL must be used to complete this sheet

Centre number:

Please write your **name** below,

then write your six digit Candidate number in the boxes and shade the number in the grid on the right in PENCIL.

0 1 2 3 4 5 6 7 8 9

Test date (shade ONE box for the day, ONE box for the month and ONE box for the year):

Day: 01 02 03 04 05 06 07 08 09 10 11 12 13 14 15 16 17 18 19 20 21 22 23 24 25 26 27 28 29 30 31

Month: 01 02 03 04 05 06 07 08 09 10 11 12 Last 2 digits of the Year: 00 01 02 03 04 05 06 07 08 09

IELTS Listening Answer Sheet

		✓ ✗			✓ ✗
1		1	21		21
2		2	22		22
3		3	23		23
4		4	24		24
5		5	25		25
6		6	26		26
7		7	27		27
8		8	28		28
9		9	29		29
10		10	30		30
11		11	31		31
12		12	32		32
13		13	33		33
14		14	34		34
15		15	35		35
16		16	36		36
17		17	37		37
18		18	38		38
19		19	39		39
20		20	40		40

Checker's Initials | Marker's Initials | Band Score | Listening Total

IELTS L-R v4.0 DP500/392

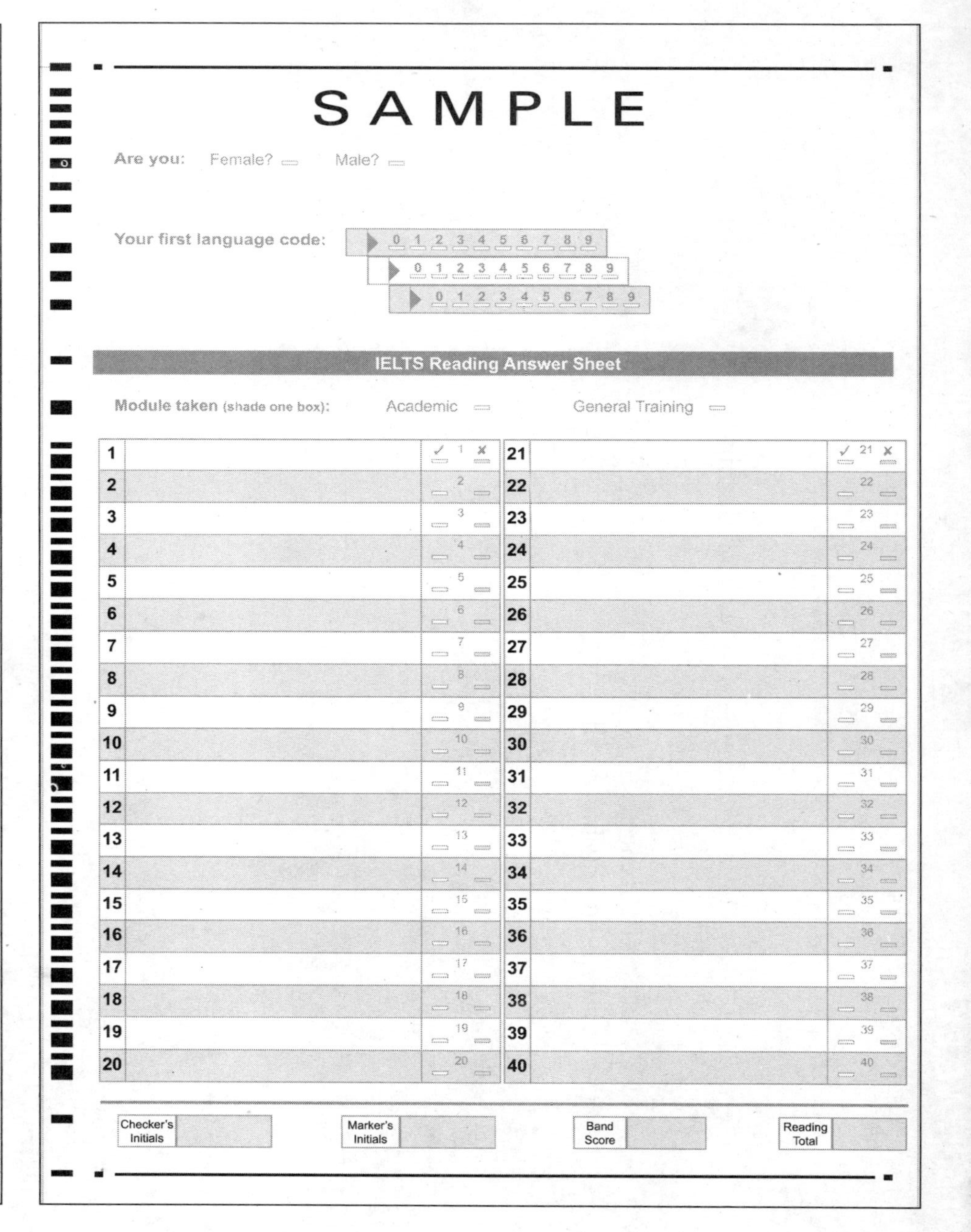

SAMPLE

Are you: Female? Male?

Your first language code: 0 1 2 3 4 5 6 7 8 9

IELTS Reading Answer Sheet

Module taken (shade one box): Academic General Training

		✓ ✗			✓ ✗
1		1	21		21
2		2	22		22
3		3	23		23
4		4	24		24
5		5	25		25
6		6	26		26
7		7	27		27
8		8	28		28
9		9	29		29
10		10	30		30
11		11	31		31
12		12	32		32
13		13	33		33
14		14	34		34
15		15	35		35
16		16	36		36
17		17	37		37
18		18	38		38
19		19	39		39
20		20	40		40

Checker's Initials | Marker's Initials | Band Score | Reading Total

© UCLES 2006 PHOTOCOPIABLE